The Cultivation of Hemp

Botany, Varieties, Cultivation
and Harvesting

Dr. Iván Bócsa
Michael Karus

Sebastopol, California

THE CULTIVATION OF HEMP: BOTANY, VARIETIES, CULTIVATION AND HARVESTING
© 1998 by HEMPTECH. All rights reserved. This book may not be reproduced in whole or in part, in any form whatsoever, without written permission from the publisher, except for brief quotations embodied in critical articles or reviews.

Translated by Chris Filben
Technical editors: Dr. Gero Leson, Petra Pless, Dr. Dave West
Cover design: Catherine Greco
Interior design & typesetting: Sara Patton
Editor: Sonia Nordenson
Printed in the U.S. by Vaughan Printing
ISBN 1-886874-03-4

Original German edition: *Der Hanfanbau:*
Botanik, Sorten, Anbau und Ernte © 1997
by C. F. Müller Verlag, Hüthig GmbH, Heidelberg
English-language Web site: www.huthig.com
German-language Web site: www.huethig.de
ISBN 3-7880-7568-6

If your local bookstore or catalog is currently out of *The Cultivation of Hemp*, call (800) 265-4367 or (419) 281-1802 twenty-four hours a day, seven days a week, or email: orders@bookmaster.com. Visa and MasterCard accepted.

HEMPTECH
P.O. BOX 1716 ▪ Sebastopol, CA 95473
(707) 823-2800 ▪ *fax:* (707) 823-2424
email: info@hemptech.com
Web site: www.hemptech.com

HEMPTECH
The Industrial Hemp Information Network

Contents

■ ■ ■ ■ ■ ■

Publisher's Preface

Because Hungary and Germany have been leaders in the revival of industrial hemp, much important hemp farming and market research has been published in either Hungarian or German. In 1996, we at HEMPTECH learned that Dr. Iván Bócsa, the eminent Hungarian plant breeder, had collaborated with the nova Institute's Michael Karus, Germany's leading hemp consultant, to publish a book about the cultivation of hemp. We are very pleased to publish this book for English-speaking readers across the globe.

A few caveats: As a highly trained breeder, Dr. Bócsa does not hesitate to recommend the use of agricultural chemicals in specific situations. Farmers should note that British, Canadian, Dutch, French, and German farmers have found no economic need to use pesticides on their commercial hemp crops. Dr. Bócsa's experience shows, however, that hemp growers cannot ignore potential pest challenges, especially in those regions where hemp acreage increases significantly.

HEMPTECH is proud to add *The Cultivation of Hemp* to its growing list of hemp publications, which includes *Industrial Hemp* by HEMPTECH; *Hemp Horizons* by John W. Roulac, and *Bioresource Hemp* by nova Institute. Our Web site (www.hemptech.com) offers an up-to-date hemp news service as well as an international directory for hemp products, seed sources, and harvesting machinery.

American readers should note that the United States government has not issued any commercial hemp-growing permits since the 1950s. We encourage readers to contact their government representatives about the opportunities in hemp farming. A good question might be: "Why are farmers in twenty-nine countries growing, processing, and exporting to the United States hemp fiber, seed, oil, and finished goods, while American farmers are restricted from participating in this ecologically sound agricultural industry?"

John W. Roulac
President, HEMPTECH
Sebastopol, California

Introduction

■ ■ ■ ■ ■ ■ ■ ■ ■

In western Europe, interest in the hemp plant and its abundance of possible uses has increased rapidly in recent years. In fact, since 1989 the cultivation of hemp in the European Union has increased by nearly 700%.

During the twentieth century, in every western European country except France, hemp cultivation was given up for decades. Hemp, the ancient European crop, fell into oblivion there, along with much knowledge about its varieties, location requirements, cultivation, and harvesting. Recently, there has been a resurgence in the development of test crops as well as new commercial uses of hemp in England, the Netherlands, Austria, and, more recently, Germany. This development has resulted in a renewed demand for the lost knowledge of hemp cultivation.

In eastern Europe, the cultivation and researching of hemp never came to a standstill. The same practical, tried-and-true knowledge was passed on, and expanded upon through modern methods. New varieties were developed, and new production records achieved. The purpose of this book is to share this extensive knowledge and practical experience gleaned from eastern Europe and, in particular, Hungary.

It was a historical stroke of luck that the well-known Hungarian hemp breeder Professor Iván Bócsa, one of the world's leading authorities on the cultivation of hemp, was interested in summarizing his decades of experience in a book written for hemp aficionados and laypersons alike. The translation of the text from Hungarian to German was certainly interesting but proved to be quite challenging, as did the adoption of Hungarian practices to German and western European conditions. The attempt to expand current knowledge with regard to western European processing and ecology proved to be an equally formidable task.

We believe that this book will significantly contribute to the reintegration of the hemp plant into the worldwide agricultural industry. Special thanks are due to Daike Lohmayer (nova Institute) for her editorial work and valuable contributions. Thanks are also due to Dr. Hayo van der Werf (International Hemp Association), Professor Dr. Karin Krupinska (University of Cologne), and Renate Huppertz (nova Institute) for their critical proofing; to Willy Leson, Leo Leson, and Dr. Gero Leson (Consolidated Growers and Processors) for editorial work, proofreading, and organizational support; and, finally, to Alexander Schaefer for typesetting.

We are very proud to now present the English version of the successful German edition. We would especially like to thank John Roulac of HEMPTECH, Sonia Nordenson, Petra Pless, Dr. Gero Leson, Daike Lohmayer, and Dr. Dave West for their outstanding translation and publishing work.

– Michael Karus, Hürth, Germany
October 15, 1997

Chapter 1

■ ■ ■ ■ ■ ■ ■

Hemp's Historical Significance

In the past, hemp had an importance equal to that of today's petrochemicals. Wars were waged because control over hemp resources was prerequisite to a display of economic power. Before coal, oil, and gas could be mined on an industrial scale in Europe and the importing of bulk goods such as cotton, jute, sisal, and ramie from overseas markets was possible, hemp processing was a key industry in Europe. (Hingst/Mackwitz 1996)

The hemp plant (*Cannabis sativa*) is one of the oldest and most versatile plants known to man. Hemp has been cultivated over a period of many centuries in almost every European country. It once represented a significant raw material source for the production of rope, canvas, textiles, paper, and oil products.

The historical significance of hemp is primarily based upon the use of hemp fiber as a technical textile. Hemp fiber, due to its extreme durability and weatherproof qualities, was virtually predestined for technical applications, and its use has been closely linked to technological advances throughout history.

In ancient times, people used plant stalks for binding. Strips of leather were used as binding materials until the development of hemp ropes. In Egypt, papyrus fibers were also used to a certain extent. The first ropes made from hemp fiber were spun in China about 2800 BC. These new Chinese ropes proved to be extremely durable and weatherproof. They were used for a wide range of purposes, but quickly gained prominence in the shipping industry as rigging for sailing ships.

The world's first paper was made from hemp fibers. A paper sample discovered near Xian (China) which dates to 140 to 87 BC contains hemp fibers and is probably the oldest paper in the world. It was produced with a floating sieve, from which the dip sieve was developed. In China, bast from the mulberry tree eventually became the most important raw material for paper. It wasn't until the thirteenth century that the technique of paper production made its way from the Near East to Italy, from where it was subsequently disseminated throughout the rest of Europe. In the fourteenth century, the art of paper production was introduced in Germany.

The utilization of hemp fibers for clothing and consumer textiles also has a long history. In 1927, Otto Heuser wrote that, according to Dewey, "the use of hemp fibers has been handed down to us through time from China. Supposedly a reference can be found in a Chinese work from the Sung Dynasty dating back to about 500 BC, stating that as early as 2800 BC, Emperor Shen Nung had taught the Chinese people how to cultivate 'ma' (hemp) and make clothing from this material."

Further references to hemp textiles during this period were cited by Abel in 1980: "In 1972, pieces of fabric were found along with other artifacts in a tomb from the Chou Dynasty (1122 to 249 BC). Analyses revealed that the fabric was made from hemp fibers. It is the oldest preserved hemp product in existence."

According to Körber-Grohne, the earliest discovery in Europe dates back to the beginning of the pre-Roman Iron Age (Hallstatt Period, 800–400 BC). One of the finds is from the burial mound of the Celtic Prince of Hochdorf near Stuttgart, Germany, dating to 500 BC. Materials woven from hemp bast were very important during this period. They were not manufactured from the processed pure fibers but rather the bark was stripped from the stalk in small, thin strips, spun, and woven in a variety of patterns. The earliest woven fabric made entirely from processed hemp fibers was discovered in the tomb of the Merovingian queen Arnegunde, who was buried between 565 and 570 AD at St. Denis Cathedral in Paris.

In the seventeenth century, during the golden age of sailing ships, hemp had its heyday in Europe. Nearly all sailing canvas, as well as the majority of rigging, ropes, nets, flags—even the uniforms worn by sailors — were made from hemp. This was primarily due to the strength of hemp fibers and their resistance to wear in wet conditions. Both trade and wartime activities were dependent upon hemp. Fifty to one hundred metric tons (55 to 110 short tons) of hemp fiber were required as basic equipment on ships, and had to be replaced every one to two years.

Until the eighteenth century, field- and water-retted mechanically processed hemp fibers, along with flax, nettle, and wool, were the raw materials of the European textile industry. Because of its more coarse and nonhomogeneous fiber bundles, hemp was primarily used in the production of outer garments and work clothes. Flax and nettle were used for finer fabrics, and wool for warmer clothing.

The Decline of Hemp Cultivation in Germany

In the eighteenth century, cotton-spinning machines were mechanized with the advent of the "spinning jenny," making the processing of cotton fibers significantly easier and less costly. From then on, the local plant fibers, whose processing remained labor-intensive, were increasingly replaced on the textile market. Until the beginning of the twentieth century, the most important technical applications for hemp fibers were the manufacturing of ropes, twine, and canvas, the latter providing the most durable technical fabric with a multitude of uses. However, by the nineteenth century hemp also lost importance in the technical arena:

Hemp supplies one of the most durable and longest-lasting long fibers which once fashioned an important raw material for spinning mills throughout the world. During the second half of the last century, hemp fiber was increasingly replaced by other fibers that are not nearly as durable but less costly than hemp. Above all, Indian jute, a coarse and not very durable fiber that, due to the low labor costs in India, can be brought to market in mass quantities. There are also a number of hard fibers, such as sisal and Manila hemp, that have begun to successfully compete with hemp fibers, thus leading to a further reduction in hemp cultivation. Hemp is not easily replaced in the production of fabrics that require extreme durability or resistance to wear in wet con-

ditions, such as canvas or certain types of rope. On the other hand, as a result of tropical competitors, jute has easily and almost entirely replaced hemp in the production of pack cloth and sacks as well as most types of low-quality rope and twine. (Heuser 1927)

In addition to receiving stiff competition from jute, sisal, and Manila hemp, German hemp was further relegated as a result of less expensive hemp imports from Russia. By 1910, hemp cultivation in Germany had nearly come to a standstill.

The Rediscovery of Hemp During World Wars I and II

In Germany, hemp became a war profiteer during both World Wars. Cut off from imports of fibers like cotton, jute, sisal, and ramie, the Germans reconsidered the use of hemp. They improved cultivation techniques and harvesting methods while expanding the uses of the plant. The so-called "cottonization" process provided a short-fiber, high-quality cotton substitute made from hemp long fibers. In the 1920s, Germany considered replacing all cotton imports with hemp fibers, which would have required roughly 1,000,000 hectares (2,500,000 acres) of arable land. This scenario undoubtedly arose primarily from Germany's endeavors to achieve self-sufficiency, but the importance placed upon hemp at that time is nonetheless apparent. Today, ecological points of view and new perspectives for European agriculture are the driving forces behind the reappearance of hemp in the fields.

During the final years of World War II, roughly 21,000 hectares (52,000 acres) of hemp were cultivated in Germany, meeting approximately 20% of the country's requirements. The remainder was imported, primarily from Italy. Near the end of the war, German cotton gins processed more "cottonized" hemp than they did cotton.

From a technical point of view, the various German innovations in fiber separation and processing technologies at the beginning of the twentieth century are of considerable current interest. These techniques were developed and tested in order to reduce the cost of fibers and simultaneously expand the spectrum of applications for hemp fibers; this resulted, for instance, in the use of hemp fibers as reinforcement in the construction of freeways.

Chapter 2

■ ■ ■ ■ ■ ■ ■

Hemp Cultivation Today

By the end of World War II, hemp was being forgotten once again. The economical cotton imports returned to assert their presence in the garment industry. Additionally, marked advances were made with synthetic fibers, which especially took over in the technical textile sector.

Accordingly, hemp cultivation in West Germany, as in most other western European countries, rapidly dwindled to the point of insignificance. In former eastern Germany, hemp cultivation continued until the end of the 1960s.

Hemp research was also in a steady decline. However, it is important to note that up to the 1950s and 1960s new varieties of hemp with low THC content and high fiber yields were bred in western Germany, and subsequently to a certain extent in France.

The cultivation of fiber hemp in western Germany was decisively banned with the Amendment to the German Drug Law (BtMG) on January 1, 1982. Hemp cultivation was prohibited regardless of THC content or intended use. The only exceptions were: (1) approval for research purposes or (2) if sufficient public interest to justify cultivation could be proven. Until 1996, this legal framework prevented any commercial cultivation of hemp. As a result of the ban on cultivation, interest in the conducting of research on agricultural hemp faded away.

Martin Butter, the last hemp farmer in West Germany, was forced to give up hemp cultivation in the Swabian Mountains after the Amendment to the German Drug Law was introduced. At that time, fifteen people worked on his farm, cultivating and processing 150 hectares (370 acres) of hemp. The fibers were delivered to rope-making factories or used in the manufacture of cigarette paper. The hurds were used to make particleboard and loose, granular insulation for homes. Farmers used the leaves as animal bedding in their stables. Butter received DM 800,000 (U.S. $450,000) as compensation from the government and in turn agreed not to file charges at the European High Court, for at that time the European Union (EU) actually permitted the farming of low-THC hemp varieties (compare to the text in the box at right).

Hemp's Possibilities Were Overlooked

In the 1980s, when resources such as flax and canola (rape) were beginning to be recognized as options for both the European agrarian economy and the consumer market, hemp was simply overlooked. The reason: the drugs marijuana and hashish can be derived from certain varieties of cannabis. For decades, specific varieties of fiber hemp have existed that cannot be used as drugs because they contain such minute quantities of THC (the psychoactive substance). But hemp was ignored in many countries and farming was even banned in certain areas, regardless of THC content or lack thereof. Agricultural researchers and the agricultural industry as a whole turned a blind eye to hemp. Hemp industries continued only in France (the sole western European country) and eastern Europe (see below). The production of coarse technical textiles for such purposes as the making of tarps and uniforms for the Russian market was predominant in eastern Europe until the breakup of the Soviet Union and the Warsaw Pact, while in western Europe hemp was cultivated on small parcels of land for use in the production of high-quality specialty pulp, primarily for the cigarette paper industry.

Cultivation Continues in Central and Eastern Europe

In contrast to what happened in western European countries, hemp cultivation was never abandoned in central and eastern Europe. Following World War II, the Soviet Union cultivated more hemp than any other country (140,000 hectares or 345,000 acres). By 1990, total hemp acreage had decreased to 40,000 hectares (100,000 acres). The yields were very low, and the fiber quality too poor for textiles. Romania also cultivates considerable amounts of hemp, and has a highly

International Growing Bans

The first bans on the cultivation of hemp were introduced in the United States within the framework of marijuana prohibition. During the 1930s, hemp cultivation was still prevalent in Kentucky, Minnesota, Virginia, and Wisconsin, until the Marihuana Tax Act of 1937 restricted cultivation. A number of other countries (such as Australia and Canada) then followed suit and also banned hemp cultivation. In 1970, cannabis restrictions were tightened by the Comprehensive Drug Abuse Prevention & Control Act, 91-513/1970.

Since 1970, the cultivation of low-THC varieties of hemp in the EU has been regulated by strict guidelines. Only varieties included in the European Union List of Varieties can be used. In some EU countries, hemp cultivation, regardless of THC content, was banned through the enactment of national laws. As recently as 1995, the growing of hemp was still forbidden in Germany and Italy. But by 1996, Italy was the only country in the European Union where hemp cultivation remained illegal. It is interesting to note that Italy was previously the home of a hemp textile industry of particularly high quality. As we go to press, Canada is preparing to issue in 1998 its first commercial hemp-growing licenses in fifty years.

developed hemp industry. However, the yields per hectare or acre are quite low. Hemp cultivation in Poland and in today's Yugoslavia is also significant, and the quality of southern varieties is as good as that of hemp cultivated in Hungary or Romania.

The highest average yield in Europe of 9 tons (metric) of hemp stalks per hectare (8,000 pounds per acre) was attained in Hungary in 1989 on 6,000 hectares (15,000 acres). In 1990, Hungary produced 5,000,000 square meters (54,000,000 square feet) of polypropylene hemp tarps for the Soviet Union (the qualities of hemp and polypropylene complement one another in such items as bags and canopies). In Hungary, a country with very few forests, the hurds, byproducts of fiber production (roughly 40,000–50,000 metric tons per year, or 44,000–55,000 short tons per year), were used as a wood substitute and as fuel, and also made their way into the furniture industry.

Hungary's reduction in hemp cultivation during the last few decades is due in part to the fact that per-hectare (or per-acre) stalk and fiber yield has increased, while demand has remained steady. Hemp seeds are cultivated in other regions than those where fiber hemp is cultivated, and different techniques are used. Hungary previously exported 500–800 metric tons (550–880 short tons) of hemp seeds in a single year. A separate line of production was subsequently developed that secured foreign markets and increased independence from the fluctuations of local demand for fiber hemp seeds. The production of hemp seeds underwent a severe crisis because antiquated, nonmechanized methods of harvesting remained prevalent. However, previous markets can be regained thanks to the development of new techniques and partially mechanized harvesting methods.

Hemp cultivation abruptly decreased in central and eastern Europe following the disintegration of the Soviet Union in 1991. Many countries in those regions had exported enormous crop quantities to the Soviet Union in the form of technical textiles. Now hemp cultivation in Hungary and Romania fell to less than 1,000 hectares (2,500 acres).

Whether by chance or by luck, the worldwide rediscovery of hemp was initiated before the industrial infrastructure collapsed entirely in those countries. The renewed interest in hemp products (particularly hemp clothing) in the United States and western Europe may well prove to be the salvation of important processing methods. Central and eastern Europe's total hemp cultivation is expected to increase to a few thousand hectares (several thousand acres) in the coming years.

A Worldwide Rediscovery

The rediscovery of hemp as a universal resource occurred in the early 1990s in the United States, Europe, and Asia. This is particularly interesting because, previously, hemp had steadily been losing its perceived importance worldwide. For example, in 1993, hemp's share of international fiber production in the textile industry amounted to 0.3%. One of the triggers to this global rediscovery of hemp was a book by Jack Herer entitled *Hemp and the Marijuana Conspiracy: The Emperor Wears No Clothes*. The 1993 German publication was entitled *Die Wiederentdeckung der Nutzpflanze Hanf-Cannabis-Marihuana* (The

Rediscovery of Hemp-Cannabis-Marijuana), and the expanded German edition contained a Europe-specific section (Brökers) and an academic addendum (Karus and others). The comparatively high EU subsidies for hemp cultivation and government support for the development of new technical applications for flax and hemp have likewise contributed to hemp's rediscovery.

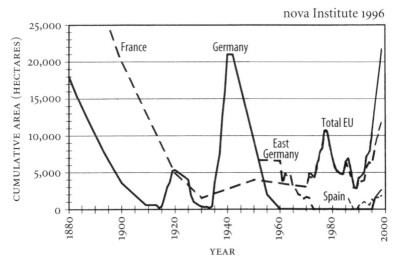

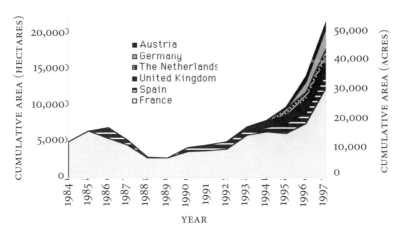

Figure 1A: Hemp cultivation in western Europe

After Spain readopted hemp cultivation in 1986 for specialty paper production, Great Britain followed suit in 1993, the Netherlands in 1994, and Austria in 1995. Germany eventually readopted hemp cultivation in 1996, and Canada in 1998. Since 1989, hemp acreage in the EU has undergone an eightfold increase, from 2,762 hectares to 21,700 hectares (6,825 acres to 53,620 acres; also compare Tables 1A and 1B).

Worldwide Data for Hemp Cultivation

Tables 1A and 1B provide a comprehensive overview of worldwide hemp acreage. The data were compiled from numerous sources. It is particularly difficult to obtain reliable data for Asia because the term "hemp" is used by the FAO (Food and Agriculture Organization) to describe a variety of exotic fiber plants (sisal, ramie, Manila hemp, sunn hemp, etc.), thus making it virtually impossible to determine the actual amount of hemp cultivated in Asia. In India, the most frequently cited statistics indicate that hemp acreage amounts to roughly 130,000 hectares (320,000 acres), although fiber hemp is said to be not currently cultivated in India. For these reasons, data for Asian countries listed in Table 1B were not obtained from FAO but rather from direct sources in the respective countries.

Recent data for eastern European acreage can only be estimated because no reliable statistical data are available, due to political and economic changes as well as the disintegration of certain eastern European countries.

Table 1A: Hemp cultivation in France, Italy, and Germany
in hectares (acres) from 1840 to 1997

Year	France	Italy	Germany
1840	176,000 (435,000)		
1850	125,000 (310,000)		
1860	100,000 (250,000)		
1878	21,238 (52,479)		
1880	40,000 (100,000)		
1883	15,308 (37,826)		
1893	7,934 (19,604)		
1900	20,000 (50,000)	3,571 (8,824)	
1909	600 (1,483)		
1910	600 (1,483)		
1912	13,870 (34,273)		
1913	12,546 (30,779)	86,600 (214,000)	600 (1,483)
1914	87,200 (215,500)	200 (494)	
1915	8,976 (22,180)	88,200 (217,900)	417 (1,030)
1916	8,750 (21,621)	86,200 (213,000)	1,600 (3,950)
1917	8,195 (20,250)	90,000 (222,000)	3,000 (7,400)
1918	9,152 (22,615)	91,000 (225,000)	3,650 (9,020)
1919	7,631 (18,851)	91,500 (226,000)	4,250 (10,500)
1920	7,147 (17,660)	95,300 (235,500)	5,350 (13,220)
1921	84,800 (209,500)	3,570 (8,820)	
1922	5,248 (12,968)	53,400 (132,000)	2,300 (5,680)
1923	4,722 (11,668)	67,950 (167,900)	2,590 (6,400)
1924	4,483 (11,077)	80,000 (197,500)	3,651 (9,022)
1925	5,342 (13,200)	3,570 (8,821)	
1926	4,926 (12,172)	105,130 (259,780)	1,908 (4,715)
1927	1,000 (2,500)		
1928	800 (2,000)		
1929	528 (1,305)		
1930	1,500 (3,700)	378 (934)	
1931	302 (746)		
1932	320 (791)		
1933	210 (519)		
1934	366 (904)		
1935	3,636 (8,985)		
1936	5,630 (13,910)		
1937	7,510 (18,560)		

– continued on next page

Table 1A (*continued*): Hemp cultivation in France, Italy, and
Germany in hectares (acres) from 1840 to 1997

Year	France	Italy	Germany
1938	12,684 (31,342)		
1939	15,848 (39,160)		
1940	21,000 (51,900)		
1941	21,000 (51,900)		
1942	21,000 (51,900)		
1950	4,000 (9,900)		
1955	2,000 (4,900)		
1960	60 (148)		
1970	3,030 (7,490)	36,466 (90,107)	2,296 (5,673)
1971	4,350 (10,750)	407 (1,006)	6,047 (14,942)
1972	4,205 (10,391)	350 (865)	9,779 (24,164)
1973	4,113 (10,163)	162 (400)	43 (106)
1974	5,560 (13,739)	186 (460)	30 (74)
1975	7,597 (18,722)	224 (554)	44 (109)
1976	7,788 (19,244)	321 (793)	27 (67)
1977	10,595 (26,180)	200 (494)	31 (77)
1978	10,498 (25,941)	325 (803)	33 (82)
1979	8,601 (21,253)	41 (101)	
1980	6,833 (16,884)	19 (47)	
1981	5,484 (13,551)	18 (44)	27 (67)
1982	5,146 (12,716)	0	0
1983	4,867 (12,026)	0	0
1984	5,000 (12,355)	0	0
1985	6,457 (15,955)	0	0
1986	5,481 (13,544)	0	0
1987	4,500 (11,120)	0	0
1988	2,750 (6,795)	0	0
1989	2,750 (6,795)	0	0
1990	3,657 (9,036)	0	0
1991	3,794 (9,375)	0	0
1992	3,946 (9751)	0	0
1993	5,862 (14,485)	0	0
1994	6,352 (15,696)	0	0
1995	6,143 (15,179)	0	1 (2)
1996	7,500 (18,500)	0	1,423 (3,516)
1997	12,500 (30,900)	0	2,812 (6,948)

Source: nova 1997 et al

Table 1B: Worldwide cultivation of hemp in hectares (acres) between 1976 and 1997

Year	Netherlands	Spain	England	Austria
1976				
1977				
1978				
1979	2 (5)			
1980	22 (54)			
1981	18 (45)			
1982	0			
1983	0			
1984	0			
1985	0			
1986	0	1,420 (3,510)		
1987	0	707 (1,746)		
1988	0	100 (247)		
1989	0	12 (30)		
1990	0	484 (1195)		
1991	0	720 (1,780)		
1992	0	1,050 (2,590)		
1993	1 (2)	786 (1,941)	407 (1,005)	
1994	137 (338)	547 (1,351)	872 (2,154)	
1995	933 (2,305)	1,371 (3,386)	1,119 (2,764)	160 (395)
1996	1,100 (2,700)	1,500 (3,700)	1,750 (4,320)	750 (1,850)
1997	1,400 (3,500)	2,000 (4,900)	2,000 (4,900)	1,000 (2,500)

Year	Switzerland	EU Total	Poland	Hungary
1976		8,136 (20096)		8,574 (21,178)
1977		10,826 (26,740)		9,902 (24,458)
1978		10,856 (26,814)		9,278 (22,917)
1979		8,644 (21,351)		9,000 (22,200)
1980		6,874 (16,979)		7,516 (18,565)
1981		5,547 (13,701)		5,310 (13,116)
1982		5,146 (12,755)		4,948 (12,222)
1983		4,867 (12,021)		5,797 (14,319)
1984		5,000 (12,400)		6,017 (14,862)
1985		6,457 (15,949)		6,511 (16,082)
1986		6,901 (17,046)		3,430 (8,472)
1987		5,207 (12,861)		6,718 (16,593)
1988		2,850 (7,040)	3,000 (7,400)	4,000 (9,900)
1989		2,762 (6,822)		2,000 (4,900)
1990	10 (2)	4,151 (10,253)		1,000 (2,500)
1991	10 (2)	4,524 (11,174)		2,000 (4,900)
1992	10 (2)	5,006 (12,365)	2,009 (4,962)	500 (1,240)
1993	12 (62)	7,068 (17,458)		1,000 (2,500)
1994	25 (62)	7,933 (19,595)		1,000 (2,500)
1995	150 (370)	9,877 (2,440)		1,500 (3,700)
1996	150 (370)	14,173 (35,007)		1,500 (3,700)
1997	150 (370)	22,000 (54,400)		900 (2,200)

– continued on next page

Table 1B (*continued*): Worldwide cultivation of hemp in
hectares (acres) between 1976 and 1997

Year	Romania	Ukraine	USSR/ Russia	Turkey
1976				
1977				
1978				
1979				
1980	35,500 (87,700)	31,600 (78,100)		
1981				
1982				
1983				
1984				
1985	46,600 (115,100)			
1986				
1987				
1988			100,000 (247,000)	4,000 (9,900)
1989	46,100 (113,900)		53,000 (130,900)	
1990	16,600 (41,000)	10,200 (25,200)		
1991	14,000 (34,600)		35,000 (86,500)	
1992	9,900 (24,500)			
1993	2,300 (5,700)	5,700 (14,100)	35,000 (86,500)	
1994	1,900 (4,700)	4,000 (9,900)	35,000 (86,500)	
1995	1,000 (2,500)	4,000 (9,900)		
1996	1,000 (2,500)	4,000 (9,900)		
1997	2,000 (5,000)	3,500 (8,700)		

Year	China	India	North Korea	Pakistan
1976				
1977				
1978				
1979				
1980				
1981				
1982				
1983				
1984				
1985				
1986	28,667 (70,807)			
1987	37,400 (92,400)			
1988	23,533 (63,067)	90,000 (22,200)	15,000 (37,100)	8,000 (19,800)
1989				
1990				
1991		77,000 (190,200)		
1992				
1993				
1994				
1995	23,810 (58,800)			
1996				
1997				

Chapter 3

■ ■ ■ ■ ■ ■ ■

Hemp's Origin and Botany

Hemp's native habitat is Central Asia, where to this day it still grows wild in Iran, Afghanistan, the southern part of Kazakstan, and some parts of southern Siberia. From these areas, it has spread to the East as well as to the West. Hemp was indigenous to the East long before it was introduced to Europe. In China, written references dating back to 500 BC and to the Sung Dynasty point to an even earlier period of hemp cultivation. No written sources exist to document the distribution of hemp cultivation in India. Nevertheless, it is assumed that hemp sucessfully spread over the mountain passes to the subcontinent, either from China or from Central Asia.

The Greeks became acquainted with hemp relatively late; Herodotus mentioned it as a "new plant"; the Roman Plinius merely acknowledged hemp as a medicinal plant. Hemp seeds were not found in the Pharaohs' tombs, nor was hemp fiber used for mummy cloths. It is thus presumed that hemp was unfamiliar to ancient Egypt as well as to the early cultures of the Near East. Hemp was not found in the Swiss pile dwellings, either.

Hemp spread from its native habitat toward the West in two directions. One route led through the Russian lowland plains to Scandinavia, extending to Poland, Germany, and the Baltic region. This distribution area included the Carpathian Mountains in the south, and went as far as the Danube river delta. This is where the northern and central Russian geographical race of hemp originated.

The other route led through Asia Minor to the Mediterranean countries and into the provinces of the Roman Empire (Illyria, Gallia, and Hispania). From here, the southern (Mediterranean) ecological form group originated, which encompassed southern Russia, Romania,

Hungary, Serbia, Italy, and Spain. In central and northern Europe, hemp was introduced by the Slavs. Presumably, distribution and cultivation did not occur to a large extent until the Middle Ages. The oldest written reference to hemp dates from the eighth century, and is found in Charlemagne's well-known *Capitulare de Villis.* In the north, Otto von Bamberg made the first known reference to hemp in a Pomeranian charter dating from 1124, referring to it as "one of the plants grown abundantly by the Slavs." The predecessors of today's southern (Mediterranean) geographical race developed along the southern border.

Hemp has been cultivated for centuries in Hungary. Evidence supporting this can be found in a customs tariff from Estergom dating to 1198. This charter, written in Latin, compared hemp to the flax plant and dictated that a customs duty be paid upon sale or delivery. It can safely be assumed that hemp was cultivated in the Hungarian Carpathian Basin as early as the tenth century, because archeologists have found clothing and other utensils made from hemp that date to 900 AD. "Hungary," in this case, encompasses the entire Carpathian Basin, which is equivalent to the present Slovakia, Romania, and a section of Serbia, where hemp cultivation was probably introduced by the Slavs.

During the Middle Ages and in modern times, hemp was cultivated in Europe over extensive areas. In almost every community, the plant was grown on the best fields, which were located near the center of the village. Evidence of this can still be found in names of agricultural land. Accordingly, an extensive home and manufacturing industry developed around 1850 in central and southeastern Europe. With the development of larger hemp factories and spinning mills at the end of the nineteenth century, hemp farming was carried out on a large scale. Hungary, for example, had extended its hemp acreage to 80,000 hectares (200,000 acres) by 1878. Of the fiber-processing plants that were then started, some are still in business today.

After World War I, the total acreage in western Europe and central southeastern Europe decreased dramatically. This occurred for several reasons: the changing territorial connections of the hemp factories; the distances between the hemp-growing regions and the processing factories; and, not least, the reduction in demand for naval use. During World War II, demand and acreage increased again, but subsequently decreased after the war. Thereafter, the only remaining hemp-growing regions of significance were in Asia, the former Soviet Union, and the eastern European countries.

In the era of the Austro-Hungarian monarchy, statistical data on hemp was collected starting in the year 1850. But, with the exception of France, hemp has not been mentioned in western European statistical almanacs since the 1960s. The cultivation of hemp was banned in both Germany and Italy in 1982. In Germany, the ban on the cultivation of fiber hemp was finally lifted in 1996 (refer to Chapter 2).

3.1 Classification and Geographical Races

The classification of hemp has been controversial. At one time it was classified in the mulberry family (Moraceae) for its morphological and anatomic similarities; then it was classified in the nettle family (Urticaceae). Most plant-classification specialists today agree that hemp belongs to the Cannabinaceae family, a family that includes only one genus, the hemp genus (*Cannabis*) and its species (*Cannabis* spp).

However, divisions within the classification of *Cannabis* continue to be disputed. Popular opinion today is that the *Cannabis* genus has only one species, "*Cannabis sativa*," which can be subdivided into several varieties:

■ Variety: *Cannabis sativa* var. *ruderalis* JANISCH (wild hemp)

■ Variety: *Cannabis sativa* var. *vulgaris* (cultivated hemp)

■ Variety: *Cannabis sativa* var. *indica* LAM. (Indian hemp)

■ Variety: *Cannabis sativa* var. *indica* LAM. subvar. *gigantea* (giant hemp)

The most recent opinion, supported by Sisov and Serebrjakova, separates the *Cannabis* genus into two species: *Cannabis sativa* LAM. SEREBR. and *Cannabis indica* L. SEREBR. The only species of significance to us is *Cannabis sativa*. Therefore, in the following discussion we refer only to the classification of this species. Only one of the subspecies "*culta*" has economic significance, while the other, wild hemp (subspecies "*spontanea*"), is economically unimportant. According to this classification system, *Cannabis sativa* L. ssp. *culta* (cultivated hemp) cannot be systematically categorized but rather can be subdivided into geographical races or ecological form groups. It is important to note that within these groups, which are dependent on geographical origin as well as the respective country's geographical and climatic conditions, there are considerable morphological and physiological differences. Nevertheless, geographical races of various origins share a common trait: they all have the same number of chromosomes (2n=20) and interbreed with one another easily and spontaneously. The following is Sisov's and Serebrjakova's categorization of the Cannabinaceae family:

■ Species: *Cannabis indica* LAM. SEREBR. (Indian hemp)

■ Species: *Cannabis sativa* (L.) SEREBR. (common hemp)

■ Subspecies : *Cannabis sativa* (L.) SEREBR. ssp. *spontanea* (wild hemp)

■ Subspecies : *Cannabis sativa* (L.) SEREBR. ssp. c*ulta* (cultivated hemp)

a) Geographical race (ecological form group): prol. *borealis* SEREBR. (northern hemp). To this group belong the geographical races from Russia and Finland that are grown north of 60 degrees north latitude.

b) Geographical race (ecological form group): prol. *medioruthenica* SEREBR. (central Russian hemp). This race is cultivated on the majority of the total hemp acreage, generally between 50 and 60 degrees north latitude, predominantly in Russia, the Ukraine, and Poland, also in Scandinavia, and formerly in Northern Germany.

c) Geographical race (ecological form group): prol. *australis* SEREBR. (Southern or Mediterranean hemp). A hemp variety found throughout central, southeastern, and southern Europe. The Hungarian, Romanian, Italian, Spanish, Turkish, and southern Russian geographical races of hemp belong to this group. The most valuable varieties of hemp belong to this race.

d) Geographical race (ecological form group): prol. *asiatica* SEREBR. (Asian hemp). This group contains varieties cultivated in China, Japan, Thailand, and Korea.

The individual geographical races can be characterized as follows:

3.1.1 Northern Hemp

The primary characteristic of this geographical race is a very short stalk which generally does not exceed a maximum height of 1.5 meters (5 feet). This ecological form group is very similar to central Russian hemp, but northern hemp matures much sooner. The world's earliest maturing varieties of hemp belong in this group. Their global economic significance is meager. Nevertheless, the varieties within this group occupy an important position in the north because there they are the only cultivable fiber and oil plant.

3.1.2 Central Russian Hemp

This geographical race developed in the Russian plains, in Poland, and, prior to 1980, in northern Germany, primarily due to the influence of the prevailing climatic conditions between the fiftieth and sixtieth degrees of north latitude (short vegetation period, long days, high precipitation, etc.). This ecological form group occupies the largest acreage.

Important characteristics: an average vegetation period (90–110 days to seed maturation); 1.25–3.00 meters (4–9 feet) taller; a lightly branched stalk; average leaf size in comparison to leaves from southern or Asiatic geographical races, with an average of five to nine leaflets. The fiber yield is mediocre but the plants can produce large amounts of seeds. Without exception, all cultivated varieties of hemp in the European regions, in the central Asiatic section of Russia, in Poland, and formerly in northern Germany, belong to this geographical race.

3.1.3 Southern (Mediterranean) Hemp

With regard to the economic significance of hemp, this geographical race is the most important. It is predominantly cultivated in Europe below 50 degrees north latitude. The seeds do not ripen any farther north. The northern boundary for seed production coincides with the boundary of the wine-growing regions. However, fiber hemp from this geographical race can be grown considerably farther north, for example, in the regions between 50 and 60 degrees north latitude (in Russia, the Ukraine, and Poland). Since the seeds do not mature this far north, they need to be imported annually, or (as in Russia, for example) delivered from the south. The southern dioecious varieties, grown as fiber hemp in the north (e.g., in England, the Netherlands, and Northern Germany), have a higher stalk yield than do the central Russian geographical races or the transitional types from France that achieve seed maturity.

Important characteristics include a relatively long vegetation period (130–150 days to seed maturity). This geographical race has the longest stalks, measuring 2.5 to 4.5 meters (8–15 feet), which tend to branch out if the plant is freestanding. The leaves are large and the number of leaflets ranges from nine to eleven. These plants can produce high stalk yields with many fine fibers. Their seed yield is average but much lower than the central Russian varieties. Southern geographical races have the typical characteristics of a cultivated plant and, even though their seed yield does not compare to the yield of the central Russian geographical race, their international economic importance is much greater because they produce high-quality fiber, with a large yield compared to their acreage.

Southern hemp is cultivated in southern, southeastern, and central Europe. The acreage by country, from the largest to the smallest, is in the following order: Romania, Hungary, former Yugoslavia, and Bulgaria. This geographical race is also grown in the southern part of Russia (around Krasnodar) and in southern Ukraine. Most varieties from Turkey belong to this category as well. The monoecious varieties of hemp, including the French varieties, are transitional types of the southern and central Russian form groups. The early-maturing monoecious varieties are more closely related to the latter type, while the late-maturing varieties are more closely associated with the previously cited type.

3.1.4 Asiatic Hemp

The varieties within this geographical race generally develop a smaller stalk (2.5–3.0 meters, 8–10 feet). This form group tends to branch out more than others and has numerous short stalk segments. The leaves are large with a bright pastel-green coloring, and the number of leaflets is usually between nine and thirteen. The individual varieties have relatively large differences with regard to their vegetation period.

On average, the vegetation period is between 150 and 170 days, but there are a number of varieties whose seeds do not mature under central European conditions. Asiatic hemp has no economic significance in Europe, but can be used as a hybrid strain for breeding.

3.1.5 Wild Hemp

According to the Sisov-Serebrjakova classification, wild hemp is a subspecies of *Cannabis sativa*. It is mainly found in central Asia, in the Volga and Ural regions of Russia, but it is also found in Turkey, Romania, Bulgaria, and Hungary. It does not grow wild west of Hungary. The plants are very short, have many branches, and produce small seeds. Since wild hemp flowers irregularly, it is possible for it to flower simultaneously with cultivated hemp, and since both interbreed rather easily, this can result in biological degradation of the cultivated variety. The stalk is no taller than one meter (3 feet) and has an unusually large number of branches. The thousand grain weight of its seeds is 10–12 grams (0.35–0.42 ounces). Varieties with a longer stalk that occur spontaneously are not wild hemp, but rather are cultivated hemp plants growing wild. These plants have spread from the cultivated area and could be regarded as weeds. These two "wild" hemp types are neither systematically nor morphologically identical.

3.2 Botanical Description

3.2.1 Roots

Hemp has a well-developed primary root, from which numerous branched secondary roots branch out. The primary root can reach a depth of 2.0–2.5 meters (6.5–8 feet), while the secondary roots can extend to 60–80 centimeters (2–2.5 feet). The depth of the root zone primarily depends upon the physical and chemical characteristics of the soil. For example, the primary root reaches deeper in crumbly soil with a high mineral content than it does in marshy soil. In soils with high mineral content the primary root reaches down to a depth of 2 meters (6.5 feet), with most of the root mass located in the topsoil at a depth of 30–50 centimeters (12–20 inches). In marshy soil the primary root merely reaches a depth of 40–60 centimeters (1–2 feet), with the largest root mass located in the topsoil at a depth of 10 to 20 centimeters (4–8 inches).

The depth of the root zone can also depend upon the groundwater level as well as the geographical race and ecological form group. In this respect, the sex of the hemp plant is also an important issue. The root structure of male plants, because of their shorter vegetation period, is not as strong as that of female plants, which are larger and have a longer vegetation period. The development of the root structure is dependent partially upon the method of agriculture, but primarily upon the site. The root mass of a thin, densely cultivated fiber hemp plant with few branches is lower than the root structure of the same type of plant spaced farther apart. The root mass of fiber hemp contributes 8–9% of the plant's entire mass.

The root structure of hemp, in comparison to that of other economically significant plants, is poorly developed. This is particularly noticeable at the beginning of the vegetation period, when the growth of the stalk exceeds that of the roots, and it partially explains why hemp requires high amounts of nutrients and water.

3.2.2 External Stalk Morphology

■ *Height and Diameter*

Hemp is a tall, vigorous annual, and lignification is significant near the end of the growth phase. The stalk is hard and stands erect. The surface is covered with hairs, which soon lie down. They are referred to as "top" hairs or glandular hairs. The stalk is hexagonally shaped, and its surface often is vertically ribbed, especially when the plants are widely spaced.

The stalk height is dependent upon a number of factors, the most important of which are the amount of daily exposure to the sun, the variety (geographical race), the soil and available nutrients, the water supply, the spacing of the plants, and finally the sex.

Hemp is a plant with a "short-day cycle." The plant passes from the vegetative to the generative phase only if the daily exposure to sunlight is shorter than a specific maximum duration (critical day length). The duration of daylight, or the photoperiodic situation, is dependent upon geographical latitude. The stalk height and the vegetative period of the northern varieties, which are accustomed to long days, become gradually shorter toward the south. However, if the southern ecological form group, which is accustomed to shorter days, is grown in the north, the length of the stalk does not decrease up to a certain latitude. On the contrary, the stalk length increases. This is because the vegetative period changes considerably as a result of the longer days: in the north, the vegetative phase becomes very long and the seeds do not mature. The stalk length of the southern geographical race increases in central and southeastern Europe, where hemp is sown for fiber, from 1.5 to 3.0 meters (from 5 to 10 feet). On the other hand, under optimal conditions, the stalk of central Russian hemp reaches a height of 1.0–2.0 meters (3–6.5 feet) in its native environment.

Stalk height is considerably affected by soil type, physical and chemical characteristics, and nutrients or fertilizers. Hemp grown on deep loess soils with good structure and optimal water balance, or on lighter alluvial soils, develops much taller stalks than hemp cultivated on compacted cold soils with high clay content, on sandy soils that lack humus and essential nutrients, or on alkali soils that have no structure and a poor water balance.

Fertilization or the availability of nutrients has a lasting effect on stalk height. The amount of available nutrients in the soil and the application of fertilizer (particularly nitrogen) can increase the stalk length by 50–60%.

Along with climatic factors, the amount of precipitation and its distribution have a significant effect on stalk development. Hemp requires large amounts of water. A minimum of 250 millimeters (10 inches) of precipitation during the vegetative period is required for hemp to develop a substantial stalk length.

Plant spacing also plays an important role. On large plots in central and southeastern Europe, the stalks of hemp grown for seeds reach a height of 4.0–4.5 meters (13–16 feet). The growth period is rather peculiar: at first, seed hemp develops considerably more slowly than fiber hemp, but near the end of the vegetative period it grows more rapidly until it becomes taller than fiber hemp.

Finally, there is a difference between male and female plants with regard to stalk length. Male plant stalks are generally 10–15% taller than female stalks, even though the female vegetation period is six to seven weeks longer than that of the male.

Factors contributing to the stalk thickness (diameter) are similar to the factors that contribute to stalk height. The most important factor is plant spacing. The stalk diameter of fiber hemp plants grown in close proximity to one another (12–20 centimeters or 4–8 inches) can vary by 4–9 millimeters (0.2–0.4 inches). The stalk diameter increases as the available area per plant is increased. In a field where the plants are spread widely apart (2,000 square centimeters or 2 square feet per plant, which is a maximum of five plants per square meter or 0.5 plants per square foot), the stalk thickness is characteristically 20–26 millimeters (0.8–1.0 inches). The sex of the plant also determines the stalk thickness. Male plants have thinner stalks, while female stalks are shorter but thicker.

The stalk, when separated from the roots and leaves, constitutes about 65–70% of the total mass of a fully grown plant.

■ Branching
The hemp stalk tends to branch out, as leaves protrude at the branching points of the stalk segments. If fiber hemp is densely sown, the leaves fall off during the developmental phase and tiny calluses form at these points. The crop loses most of its leaves. Fiber hemp plants do not have a lot of branches. However, if the area available for one plant is expanded (to more than 2,000 square centimeters or 2 square feet per plant, which means a maximum of roughly five plants per square meter or 0.5 plants per square foot), the plants will branch out dependent upon variety, sex, and the availability of nutrients and water. Female hemp plants have a greater tendency to branch out than do males, which is manifested in their number of secondary branches as well as their average length.

3.2.3 Stalk Anatomy
Morphologically, the hemp stalk consists of wood and bast tissue. Bast tissue is composed of phloem, bast parenchyma, bast radial parenchyma, and bast fibers. Wood tissue consists of tracheids, wood paren-

chyma cells, and wood fibers, which transport primarily water and nutrients. The wood tissue forms the hurds. The wood fibers are responsible for the vertical strength of the stalk. The length of the wood fiber cells does not exceed 0.5 millimeters (0.02 inches). Wood fibers are significantly stiffer and less flexible than bast fibers.

The most important tissue for hemp cultivation and processing, the bast tissue, forms the exterior layer of the stalk. The stalks of hemp (and flax) contain bast fibers, which are not comparable to the leaf fibers or seed hairs of fiber plants cultivated in warmer regions (sisal and cotton, for example).

Bast tissue functions as the plant's transport system, carrying the products from photosynthesis from the leaves to the roots and, in exchange, carrying nutrients taken up by the roots to the leaves. In bast tissue, groups of fiber cells are formed whose outer cell walls are strengthened with thick layers of cellulose. Hemp is cultivated for these fiber cells. They are generally viewed as an integral component of the bast tissue, even though they are actually located outside, between cortex and bast. The main function of the bast fiber cells is not the transport of nutrients, but rather the reinforcement or strengthening of the stalk. These fiber cells, in contrast to wood fibers, do not provide vertical strengthening of the stalk, but rather impart tensile strength as well as break and torque resistance.

In a cross section, the fiber cells are easy to differentiate from the adjoining bast elements and the pericambium (primary cell tissue). Their length ranges from 1–10 centimeters (0.4–4 inches), the average length being about 1.5–4.0 centimeters (0.6–1.6 inches). The average thickness is 18–25 microns (0.08–0.025 Angstrom). Fiber cells rarely occur individually; they are generally incorporated in fiber bundles. The shape of the fiber bundle can have a round, oval, elliptical, or rectangular cross section.

The individual fiber cells or fundamental cells are held together in bundles by binding substances that consist mainly of pectins. The fundamental task of the fiber-processing industry is to break down the fibers by biological, mechanical, or chemical processes. The so-called technical fiber contains several fiber bundles and, besides fiber cells, contains numerous other tissue portions.

The cross-sectional shape of the individual cells is three- to seven-sided. The fibers are pressed tightly together during the secondary growth stage of the rest of the stalk, which alters their original circular shape. The ends of the fiber cells are rounded off, but often show a fork-shaped branching.

The most important component of the fiber is cellulose. In addition, fiber bundles also contain hemicellulose, pentosans, pectins, and lignins. These substances are decomposed by bacteria during retting, or broken down during chemical processing. Cellulose, on the other hand, is not decomposed as easily by these processes; therefore, relatively clean fiber bundles can be obtained by mechanical fiber processing.

If a plant is older, the fiber cells lignify. Lignification severely affects fiber quality. The substances responsible for this incrustation are collectively called lignins. These substances increase the tensile strength of the fibers, but reduce break and torque resistance as well as elasticity.

The arrangement of the fiber bundles and the dynamics of their form is characteristic for hemp. In contrast to flax, hemp forms secondary fiber bundle systems or rings throughout its growing season, in addition to the primary fibers. These secondary fiber bundles provide the plant, even when fully developed, with sufficient stability and elasticity. Since the stalks grow from the inside outwards, the older fiber bundles are located toward the outside (see Figures 1B and 1C).

The smaller secondary fiber bundles differ from the primary fiber bundles in that the individual fibers from which they are formed have a smaller diameter and are generally considerably shorter. In addition, they often adhere more strongly to the wood section of the stalk and are consequently lost in processing. The primary fibers, because of their great length and strength, are more valuable to the fiber-processing industry.

With regard to the anatomy and dynamism of fiber development, there is a difference between the two plant sexes. The "strengthening" growth in female plants is greater than in male plants, which means the female stalks are stronger than male stalks. Female plants have a considerably longer vegetative period, resulting in thicker and stronger fiber cells. Moreover, it is common for the male plants to more rapidly accumulate fibers. This developmental dynamism means that male

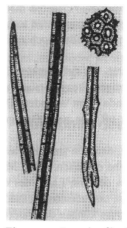

Figure 1B: Longitudinal section and cross section of a single fiber (200/1)

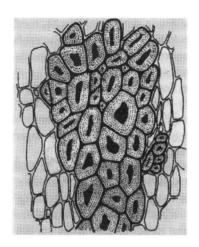

Figure 1C: Cross section of a fiber bundle (330/1)

plants contain more primary fibers than do female plants. The greater proportion of primary to secondary fiber bundles explains the relatively higher fiber content and the substantially superior quality of male plants, whereas female plants have greater tensile strength due to their higher proportion of secondary fiber bundles.

Historically, farmers pulled up the male plants on their small plots separately from the female plants, and used the male plant fibers for fine linens (bed linens, hand towels, clothing, etc.) while the female plant fibers were used to make coarser fabrics (bags, tarps, etc.). Today, quality and fineness variances between both sexes can be established with precise tests. The results in Table 2A clearly show that male plants surpass female plants in all qualitative parameters except tensile strength (test variety Kompolti). The parameters of common fiber hemp (Kompolti) in a mixed plot take a median position between both sexes. In terms of quality, its fibers are closer to male fibers than to female fibers. Table 2A also shows that most of the qualitative characteristics of a monoecious variety (Fibrimon 21) are inferior to the dioecious variety.

Fiber development is more intensive in the first phase of plant development than in the second phase. The fibers grow the most in the rapid-growth period before flowering. Growth slows during the

Table 2A: Qualitative characteristics of Kompolti (dioecious)
and Fibrimon 21 (monoecious)

Groups	Weight (%)	Fiber (%)	Flexibility (Hmm)	Torque Resistance (10^3/T)	Tensile Strength (P_{kp})	Metric Fineness (Nm)
Kompolti / Male	49	31.5	26.8	19.31	6.16	141.72
Kompolti / Female	51	29.6	18.3	13.23	7.26	109.63
Kompolti / Male+Female	100	30.4	24.3	16.78	6.04	131.23
Fibrimon 21	100	26.7	16.0	11.80	6.25	101.25

flowering stage and completely stops at the end of this period. Considering that nutrient requirements are greatest during the budding and flowering stage, it is extremely important to ensure that these requirements are met during this stage of growth.

The stalk's length/thickness ratio is important. The greater the length and the smaller the diameter, the higher the stalk's fiber quality (Table 2B).

Table 2B: Fiber content of an average variety,
depending on length/thickness ratio

Length/Thickness Ratio	230	175	170	160	155	145	120	115
Fiber Content %	24.0	20.5	19.3	18.5	18.0	17.5	16.8	15.8

Example: If the hemp plant is 115 centimeters (45 inches) tall and has a stalk diameter of 0.5 centimeters (0.2 inches), the corresponding length/thickness ratio is 230. The fiber content in this case is 24%. A plant that is 200 centimeters (79 inches) tall and 0.8 centimeters (0.3 inches) thick has the same length/thickness ratio and the same fiber content.

The optimal stalk height for industrial processing is 2.0–2.5 meters (6.5–8 feet), with a diameter of 6–10 millimeters (0.25–4 inches).

The hemp stalk bast tissue consists of 65% cellulose, 15% hemicellulose, and 4% lignin, while the hurds contain 37% cellulose, 35% hemicellulose, and 21% lignin. This equates to a total cellulose yield of 2–3 tons per hectare (1,800–2,700 pounds per acre). Fertilization, hemp varieties, and other environmental factors do not have a major influence on variations in this composition.

3.2.4 Leaves

Hemp has finely formed pinnate leaves. Each leaf contains several individual "fingers" or pinnations, called leaflets, whose number is determined by three factors. The number of leaflets is primarily characteristic for the variety. The fully developed leaf generally has five to thirteen leaflets, but those having seven to eleven leaflets are quite common. Southern geographical races of hemp normally have seven to nine leaflets. Ecological form groups from the Asiatic region show a greater number of leaflets. Leaves with eleven leaflets are more common or occur just as frequently as leaves with nine leaflets (see Figure 2).

Aside from depending upon the variety, the number of leaflets is also dependent on the age of the plant. The first true leaves have only one leaflet; the following have more, but still only three leaflets. The

Figure 2: Leaves from Chinese (Asiatic) hemp (*left*);
and from the southern variety Kompolti (*right*)

number of leaflets per leaf increases as the plant continues to develop (see Figure 3).

The number of leaflets is also determined by the position of the leaves on the stalk. During the vegetative phase, the leaves grow opposite one another on the stalk (decussate); their position in relation to one another later changes to a staggered position (alternate) at the beginning of the generative phase. The first point of such a leaf position is referred to as the GV point (see Figure 4).

The leaves are supported by the stiff leaf petioles (leaf stems). The length of the petiole can vary between 3 and 15 centimeters (1–6 inches). The leaf surface is covered with numerous top hairs and a few glandular hairs that excrete a minimal amount of THC in their resin.

Irregularly shaped leaves are very common and develop from fused leaflets, from a dual-leaf formation, or from chlorophyll defects resulting in leaflets with different colors (referred to as teratogenic formations). Leaf mass is high in relation to total plant mass; it amounts to about 24–25%. Near the end of the vegetative phase, leaf mass decreases to 8–14% because of losses due to wind as well as natural attrition.

Figure 3: Plant development from the first to the fifth set of leaves

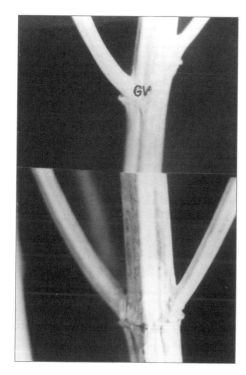

Figure 4: GV point: change from opposite to alternate growth of secondary branches

3.2.5 Flowers

Because hemp naturally occurs as a dioecious plant, male and female flowers are found on different plants. Both appear in clusters (inflorescences).

▪ *Male flowers*

The male flowers develop in pairs at the petiole nodes (intersections) of the inflorescence. The male flower consists of a simple calyx (floral sheath) with five petals that are yellowish-green and about 5 millimeters (0.2 inches) long. The petals surround five stamens that consist of anthers (pollen sacs) suspended on thin filaments (see Figure 5). The anthers have an elongated prism shape prior to maturation, and turn light-yellow following maturation. The open male flowers appear star-shaped when viewed from above.

The number of male flowers primarily depends on spacing of the plants. Fiber hemp plants having fewer branches that are thinner and

Figure 5: Development of male inflorescence
Top left: an open male flower with five stamens

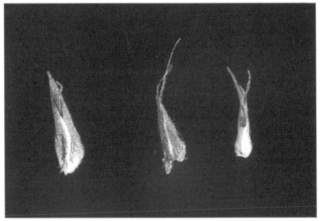

Figure 6: Female flowers with dual-forked stigmas

longer have relatively few male flowers because inflorescences develop only on the tip of the plant. One secondary branch of a widely spaced male plant has considerably more flowers than does a fiber hemp plant in a dense crop.

Hemp is a cross-pollinated, wind-pollinated plant. The pollen is dry and floury, and forms dense clouds during the flowering period.

These pollen clouds can reach an altitude of 20–30 meters (65–100 feet) and travel as far as 12 kilometers (7.5 miles). Of all cultivated plants, hemp produces the highest amount of pollen. A single male plant with a height of 3–4 meters (10–13 feet), planted in a wide row distance, can produce about 30–40 grams (1–1.5 ounces) of pollen.

▪ Female flowers

The secondary branches of the female inflorescence are very short, bearing tightly clustered female flowers. Therefore the female inflorescence is dense and club-shaped after pollination and during harvesting. The female flower consists of a green, single-leaf calyx that surrounds the hidden ovary, which has one seed (see Figure 6). Only the two thin pistils (fused style and stigma) protrude from lateral slits. The styles bear dual-forked stigmas that are initially white but later become crimson-red. The visible parts of the paired pistils are about 3–8 millimeters (0.1–0.3 inches) long.

Female flowers are very inconspicuous. Even at close range, they are hard to recognize. When pollination is late or if there is a pollen deficiency, the pistils reach an unusual length of 10–20 millimeters (0.4–0.8 inches) and turn bright white. In varieties with a high THC content, at the end of the flowering period the glandular hairs on the pistils and the surrounding calyx excrete the sticky resin used to produce hashish. The whole dried female flower is described as marijuana and is harvested from the end of flowering to initial seed growth. The THC content of cultivated varieties of hemp in Europe is so minimal that these varieties are not suited for production and use as marijuana or hashish.

▪ Flowers on monoecious plants

In some countries monoecious hemp is cultivated, and its flower structure does not differ from that of dioecious hemp. However, the location of the inflorescences is different. The male flowers are located in a whorl on the stem of the primary branches, whereas the female flowers are always situated on the tops of the primary branches. Monoecious hemp has fewer male flowers and less pollen than dioecious hemp, and also has fewer female flowers. Monoecious hemp is

Figure 7: Ideal
monoecious hemp

very diverse with regard to its flower distribution. Its appearance and
characteristics are similar to those of female plants (see Figure 7).

3.2.6 Seeds

Botanically, the fruit of a hemp plant is a small nut. However, people universally refer to hemp "seeds," which is scientifically not correct but has become custom. In this book, we will most often bow to common usage and refer to the nuts as seeds.

The tiny nut is a dry, closed achene that contains a single seed. It is surrounded by the pericarp (fruit husk), which is sometimes (especially shortly after reaching maturity) still surrounded by a leaf. Hemp seeds are either spherical or elliptical. They measure 2.5–5 millimeters (0.08–0.24 inches) in length and 2–4 millimeters (0.08–0.16 inches) in width, and their diameter ranges from 2–3.5 millimeters (0.08–0.14 inches). The thousand grain weight (TGW) of different varieties varies from 2 to 70 grams (see Figure 8). The wild form has the smallest seeds. Its TGW rarely exceeds 10–12 grams (0.35–0.42 ounces). The TGW of monoecious varieties is lower than that of dioecious varieties. The color of the seeds is light-grey to brown, and often marbled—which is not a genetic characteristic of the pericarp but rather represents a colored imprint of the surrounding bracts at harvesting. This is easy to prove, because the marbled pattern can be removed by washing or rubbing the pericarp. Thus the marbled pattern connot be regarded as a characteristic of varieties. The pericarp is very thin, with net-shaped veins; healthy seeds, not older than one or two years, have a bright luster. Old, moldy, or diseased seeds that will not germinate have a dull appearance.

The hemp seed consists of two cotyledons (seed leaves) that are rich in reserve substances, with a rootlet and a thin, undeveloped

Figure 8: *left:* Seeds from the Kompolti hemp variety
(TGW of 21 grams, 0.74 ounces); *right:* monoecious hemp
(TGW of 17 grams, 0.60 ounces)

endosperm containing starch. The cotyledons and rootlet are rich in oil. The oil content of the seeds is 30–32%. The oil dries well; it could be used for paints, lacquers, and varnishes, but is too expensive for this purpose and thus cannot compete with flax, canola, and sunflower oils. Its composition makes it excellent for use as an edible oil, particularly as a high-quality salad oil. It is also an excellent raw material for the cosmetics industry. Hemp seeds are expensive because hemp produces much less oil (about one-third) in comparison to rape (the source of canola) and to sunflowers. Its iodine value is 140–170.

Recent attempts have been made to cultivate hemp for oil in some western European countries. The southern dioecious varieties are not suitable as "oil hemp" because their seed-yielding capacity is inferior to monoecious varieties. Once again, it should be emphasized that hemp is not an "oil plant." If it is cultivated for this purpose, mid-early to early-maturing French, Polish, or Ukrainian monoecious ecological form groups should be used. The oil content of these different varieties is quite similar. Presently, hemp is not cultivated anywhere for seed and oil production alone.[1]

Authorities responsible for licensing hemp varieties regard the seed-bearing potential as a value-determining characteristic for the respective variety.

Hemp seeds rapidly lose their ability to germinate, which is at 95% in the year of harvesting and only 80% the following year; after two years the seeds are no longer suitable for sowing. The rapid decline in the germination potential can be prevented through storage at 2–3 c (36–37 f) and minimal humidity. It is worthwhile to store smaller, more valuable quantities in freezer compartments (at −12 c, 10 f), which will retain the germination potential for six to eight years with no loss in quality.

---

[1] Due to the valuable contents in hemp oil, it is expected that specialized varieties of oil hemp will be bred. These varieties will produce yields that can compete with flax oil.

3.3 The Biology of Hemp

3.3.1 Sex Characteristics

Of all our agricultural crops, hemp is one of only a few dioecious plants. The male plant bears male flowers with pollen; the female plant contains the ovary, from which the fruit later develops. The role of the male plant is simply to pollinate the female stigmas.

The occurrence of male to female plants in a normal dioecious crop is close to 1:1. However, examination of individual hemp plants in larger numbers shows that female plants slightly predominate. In a sample of several thousand plants, for every 100 male plants there are between 107 and 113 females, which equates to 42% male and 53% female plants. This proportion can vary slightly depending on geographical races, the variety, and other environmental conditions. The sex ratio within a particular variety is rather stable and cannot be significantly altered artificially, by environmental factors, or by natural selection.

Hungary cultivates dioecious plants exclusively, as do most other countries throughout the world. The monoecious variety is mainly grown in France, Russia, the Ukraine, and Poland.

Morphological, quantitative, and physiological variations exist between male and female plants. This phenomenon is referred to as sexual dimorphism; similar manifestations are found in the animal kingdom. (The morphological variations are discussed in detail in earlier sections.)

At the harvesting time for fiber hemp—the stage at which the male plants flower—the plant mass of the male and female plants is equivalent. The later the harvest, the lower the weight of the male plants, due to drying. Since the male's fiber content is greater and its fiber quality better in almost every respect than that of the female, dioecious varieties of hemp understandably are of higher quality than monoecious varieties that are similar to dioecious female plants (see Table 2A).

There are numerous physiological differences between the sexes. One primary difference is the variation in the length of the vegetative period. The male vegetative period is five to six weeks shorter than that of the female. Male plants gradually die off after completion of their biological function, the pollinating of the female plants. Female plants continue to thrive until their seeds have matured, and occasionally even longer. This difference in vegetative period also determines the variation in developmental dynamism of the sexes. Hemp breeders know that some plants are stronger and develop more rapidly than others. Normally males develop from thin, tall plants while females develop from shorter, stockier plants. Due to its briefer vegetative period, the developmental phase for male plants is more vigorous than it is for females.

The point of harvesting for densely sown fiber hemp is at the flowering stage of the male plants. For monoecious varieties, it is at the pollination point of male flowers. When the male plants flower, they are technically and biologically mature, whereas the female plants have only reached technical maturity at that point. Nonetheless, the crop is cut, because the formation of seeds in the females or the lignification and breaking of stalks after the male plants have died off would seriously affect fiber quantity and quality.

Morphological and physiological differences between the sexes are not present in monoecious hemp. As has already been mentioned, the monoecious varieties of hemp bred to date show more female characteristics and have flowers from both sexes. Male flowers from monoecious hemp flower at the same time as males from dioecious hemp; each individual plant also simultaneously produces seeds. Questions concerning cultivation and other relevant topics are discussed in detail in Chapter 4.

3.3.2 Growth and Development

The vegetative cycle of hemp plants can be divided into the following growth phases:

1. The germination stage, which lasts until the first pair of true leaves on the germinating plant reach the size of the cotyledons and are capable of photosynthesis.

2. The stage of slow growth, which lasts from the appearance of the first pair of leaves to the growth of the fifth set of leaves.

3. The rapid growth period, which lasts from the end of Stage 2 until the formation of flower buds.

4. The period between the growth of flower buds in the leaf axis and the time when the first flowers open, after which growth gradually slows.

5. Flowering, which extends from the time when the anthers on the first male flowers open and release their pollen to the time when the flowers in the upper third of the inflorescence open and their pollen is released.

6. The growth of the achene, which lasts from the initial swelling of the seed embryo to the maturation of the first seeds.

The active life and development of the hemp plant begins with germination. Necessary environmental conditions are not discussed here. Instead, we will simply concentrate on germination itself as the initial period of growth.

The rapidity and duration of the process depend upon several factors. Under normal conditions, a rootlet appears 24–28 hours after sowing. The rootlet, from which the root system later develops, immediately penetrates the soil. Soon afterward, the vegetative bud appears with both cotyledons, from which the hypocotyl (the embryonic stem) develops. The hypocotyl readily pushes up from within the soil to the surface. The cotyledons still enclose the tiny bud, which consists of two leaflets. Often half of the pericarp is still attached to the cotyledons. In the beginning the cotyledons are still yellow, but they gradually turn green.

After the cotyledons open, the small bud rapidly develops and produces the epicotyl (seedling stem). Simultaneously, the first pair of true leaves develops. The oppositely oriented pair of leaves is rotated one-quarter turn to the cotyledons (see Figure 9). Eventually, when the nutrients stored in the cotyledons are exhausted, the plant begins to live independently.

On the ninth to tenth day after germinating, the second pair of leaves appears. After another five to six days, the third pair of leaves has developed. After roughly twenty-five days, the fourth and fifth pairs emerge. The cotyledons dry out gradually and fall off the plant (see Figure 3). Around the thirtieth day, the first axial limb buds emerge in the

Figure 9: Seedling showing
the first set of leaves

second, third, and fourth nodes, which then produce more leaves. In the meantime, the stem segments become longer. The first leaves are still simple; subsequent leaves contain three leaflets, then five to seven; the leaves on adult plants predominantly have nine leaflets.

Hemp is one of our tallest-growing economic crops. This of course is true only for ecological form groups in the southern geographical race. With adequate nutrients and a steady supply of water, southern fiber hemp can reach a height of 2.5–3.0 meters (8–10 feet) in southeastern Europe. Hemp plants grown for seed that are spaced farther apart can reach a height of 4.0–4.5 meters (13–15 feet). Having only a short period of time to grow to such heights, hemp shows a very rapid growth, especially under optimal breeding conditions.

Under normal farming conditions, when the height of cultivated hemp plants is measured in short intervals up to the stage of male flowering, three clearly distinguished developmental stages can be established.

▪ In the first stage, four to five weeks after sprouting, the plants develop slowly. In this rather long period, the plants reach 15–20% of their final height.

▪ In the second stage, the hemp plants begin a radical growth phase. This is the fast-growth stage, in which the plants grow another 60% of their final height in five to six weeks. During this stage, the plants' requirements for nutrients and water are at their highest. In order to measure the rapid growth that occurs during this stage, we might test a fiber hemp plant with an average final height of about 170 centimeters (5.6 feet). In its rapid-growth stage, the plant grows 110 centimeters (3.6 feet) in forty days (i.e., 2.75 centimeters or 1 inch per day). In contrast, the average daily growth rate of a fiber hemp plant with an average final height of 220 centimeters (7.2 feet) is 3.5 centimeters (1.5 inches) during this time, and it can reach 5.0 centimeters or 2 inches per day in humid, sunny weather following rains. Some seed hemp varieties with a greater final height are said to have an average growth rate of 9–11 centimeters (3.5–4 inches) per day during this stage.

▪ In the third and final stage, growth returns to a slower rate. This period can last four to six weeks as the plants continue to grow an additional 15–20%, reaching their final height. This third stage begins when the flower buds appear in the nodes of the male plants, and ends when the flower buds reach their maximum size on the inflorescence stems. The plant stops growing when flowering ceases. The inflorescence stem simply becomes longer, and it breaks off during harvesting and defoliation. The length of the stalk does not increase after this point. These three developmental stages are depicted in Figure 10.

While the growth rate of certain varieties can vary within a particular geographical race, the growth curves are always characteristic of the varieties. Plant development is dependent upon temperature and the availability of nutrients and water.

3.3.3 Flowering

The number of days to the onset of flowering is primarily dependent on the so-called photoperiodic reaction. Hemp is a short-day plant. In other words, shorter days accelerate the beginning of flowering. For example, the central Russian geographical races flower in Hungary after only sixty days, and the same is true if the Hungarian (southern) geographical races are brought to 30 to 35 degrees north latitude (as in North Africa). Central Russian geographical races cultivated in their native region begin flowering—in comparison to what is observed in Hungary—during relatively long days. The farther south these geographical races are cultivated and the shorter the day length, the earlier the plants flower. The opposite also holds true. When southern geo-

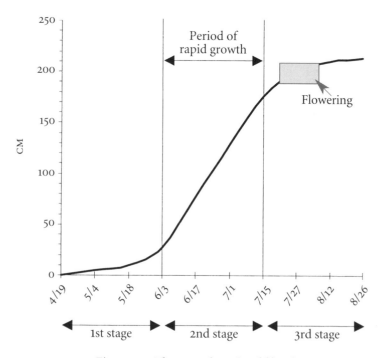

Figure 10: The growth cycle of fiber hemp

graphical races are grown in the north, flowering does not begin until the end of August or beginning of September, which prolongs the vegetative phase. The stalk yield is increased, but the seeds do not mature. Therefore, above 50 degrees north latitude, it is advisable to cultivate southern geographical races for fiber.

The subtropical and tropical ecological form groups flower in their native regions at the regular time, which means that they are adapted to the short twelve- to thirteen-hour days.

Four to six weeks pass from the emergence of buds (primordia) until full flowering. The inflorescence develops in several phases (see Figure 5). The beginning of the generative phase varies with geographical races. For example, plants from the central Russian region, under central and south European light conditions, flower as early as the end of June or beginning of July, and consequently have a much lower yield. Flowering within the same geographical race is very homogeneous

—as expressed in calendar days—and is modified only by a few days by warm weather in the preceding period. The male flowers from the southern geographical form group generally flower during the last ten days of July and continue flowering throughout the first ten days of August. High temperature reduces this period, while cooler temperatures lengthen it by a few days.

When the male plants develop flowers, or inflorescences, the flowers vary from small green-and-yellow ones to large buds and open flowers. During flowering, almost every developmental stage can be found. The flowers on the inflorescence open from bottom to top, even with widely spaced plants, on which several inflorescences are found per single plant.

Male plants and the entire associated crop reach technical maturity when one-half of the plants have yellow buds and the other half have open flowers. This is the point when harvesting begins. Male flowers open a few days before female flowers open.

Female plants flower very inconspicuously; the flowers are hardly noticeable. They can only be recognized from a close distance by detecting the fork-shaped stigma protruding from the involucre leaves; the stigma is white to yellow at first but turns red after pollination and fertilization. A few days after pollination, the underside of the pistil swells and the beginning of the fruit development is noticeable. If pollination is late or if for some reason there is a lack of pollen, the stigmas continue growing to considerable lengths (10–20 millimeters or 4–8 inches) and become vividly white in color. A stigma can be pollinated for a period of eight to fourteen days, but the period of time that the pollen is effective is shorter. In the latest tests conducted in Hungary, researchers were successful in preserving the germinating potential of pollen for more than six months by storing the pollen in liquid nitrogen at a temperature of –180 c (–292 f). It is expected that this period can be increased significantly in the future.

In the third to fourth week after pollination, the seeds begin to take shape (see Figure 11). From then on, growers speak of a late harvest, because the fiber quality rapidly deteriorates.

Flowering lasts for two to three weeks, depending on the temperature. Seed development, from the swollen female flowers to seed maturation, lasts four to six weeks.

Figure 11: Seed formation in early stage

Chapter 4

■ ■ ■ ■ ■ ■ ■

The Breeding of Hemp Varieties

4.1 Goals of Breeding

For the most part, hemp is cultivated in Europe for its fibers. The determination of breeding objectives is based on this fact. The use of hemp in the cellulose and paper industries is new except in France, where hemp has been cultivated for such purposes for nearly three decades.

By determining the objectives of cultivation, the grower starts with an analysis of the economically usable yield of a plant. It is a matter of national economic interest to obtain the highest fiber yield from a unit area.

The calculable per-unit-area fiber yield from any variety of hemp consists of two elements: the stalk yield per acre or hectare, which is the actual end product, and the amount of fiber or fiber content of the stalk (the percentage of fiber). From these two quantities, the fiber yield per acre or hectare is calculated.

The two most important objectives, therefore, are to increase the stalk yield and increase the stalk's fiber content. Previously, the hemp industry placed only moderate demands on fiber quality (fineness, divisibility, etc.). Only yarn with a quality rating of 4–5 Nm was required. Today the importance placed on fiber fineness is much greater.

However, fiber fineness is much more difficult to control through breeding than by the many promising technical processes presently being tested. These methods can improve hemp's finished fiber quality to the level of linen (mechanical processing) or even cotton (chemical-physical processing or "cottonization"; see Section 9.2). However, tensile strength is far more important than fineness, because hemp products will need to meet high standards with regard to wear and tear, strength, and ability to stretch. The aims of increasing both fiber content and tensile strength are not mutually exclusive.

Previously, more value was placed on lighter stalk color. These days, this is important only to the paper industry.

4.2 Breeding Objectives

Definitions

Dioecious: Female and male flowers on two separate plants.

Monoecious: Female and male flowers on the same plant, but on separate parts of of the stem.

Unisexual: Female dioecious plants crossed with a monoecious male variety. The first generation (F1) is 70–85% female, 10–15% monoecious, and only 1–2% male. Higher seed yields are obtained with such varieties, because of the drastic shift in sex ratio.

Hybrid cultivation: A cross between monoecious males and dioecious females in order to produce unisexual females. These are then crossed with a dioecious father. The resulting F1 generation is referred to as a hybrid.

These objectives apply to all varieties of hemp. The demands placed on hemp by the paper industry are essentially the same as those of the textile industry. High yield is of paramount importance because the fiber is a prerequisite for fine cellulose quality. From these fibers, the finest papers for cigarettes, bibles, and bank notes are produced. The question of course arises: "Which method of breeding can most easily and effectively accomplish these goals?"

Currently, a significant increase in fiber content can be obtained most rapidly with ordinary dioecious varieties. Today's breeding methods produce a total fiber content of 38–40% from plants in the breeding field trials, which results in 28–30% from industrial-hemp stalks.

Another way to increase fiber yield per acre or hectare is by increasing the stalk yield. In this respect, open-pollinated dioecious varieties have significant reserves for selection because the breeders have only selected for fiber content and have placed little weight on the stalk yield. Another method of increasing stalk yield is the use of the heterosis effect in the F1 generation. One such hybrid was developed and nationally registered in Kompolt, Hungary, in 1954. Large-scale seed production was not possible, however, because it required too much manual labor. The discovery of unisexual varieties has offered a realistic possibility for the revival of such a variety, because the hybrid seed F1 can be produced on a large-scale basis without need for mechanically removing the male organs (emasculation).

In closing, a word about monoecious hemp: its yield capacity is less than that of dioecious or unisexual varieties. It is grown at the Kompolt Research Institute only to serve as the male variety in the production of unisexual hemp.

4.3 Breeding Methods

4.3.1 Methods for Individual Plant Selection and for Breeding of Open-Pollinated Dioecious Varieties

These techniques are essential for improving quantitative as well as qualitative characteristics. Breeding for fiber content is based on the process of individual plant selection. Total stalk yield can be increased by crossbreeding or hybridization of varieties.

Hemp breeding uses three approaches to individual plant selection: (1) mass selection among half-sib families; (2) the Bredemann method; and (3) the modified Bredemann method.

Mass selection among half-sib families (the Ohio technique) is the oldest of these approaches. It is now regarded as the least successful method, because pollination cannot be controlled. This technique can aid in removing extremely weak strains from the crop, which simultaneously facilitates a slow selection improvement in increasing fiber content. However, such strains always produce male plants with low fiber content. These plants flower along with all other plants in the crop and thus negatively affect the fiber content of the entire crop. This technique does not produce a rapid increase of fiber content.

The inadequacies of the Ohio technique are eliminated in the

Bredemann method. This method allows for full control of flowering, recognizing that individual plant selection can work only when selection for fiber content is made on both male and female plants. This can be successful only if the male plant's fiber content is known prior to flowering, which requires a fiber-testing method. The procedure is as follows: the stalk of the male plants is vertically cut in half, the bark is removed, and the fiber percentage is determined by chemical processing. Only a few days are available to complete this process, because the stalk portion remaining in the ground will rapidly proceed from the budding stage to flowering. Of the hundreds of male plants tested in this manner, only those with a high fiber content are able to flower. The breeder therefore knows that the female plants harvested in the fall had been pollinated by male plants with high fiber content.

The Bredemann method can also be combined with the common methods of individual plant selection. Also, mass selection among half-sib families can be successfully implemented. In Hungary, the Bredemann method is used to increase the fiber content of hemp. Because of this technique, Hungarian varieties are some of the best in the world in fiber content.

Hemp cultivation has a rich history in Hungary, spanning the past 60–70 years. A number of outstanding breeders contributed to the introduction of Italian varieties of hemp and bred new varieties from those. These varieties were not yet bred for fiber content, and so could not meet the fiber industry's increasing demand for fiber.

Hemp cultivation for higher fiber content began in 1950–1952. The introduction of Bredemann's selection principle brought about notable success in the cultivation of dioecious hemp. In 1959, the nationally registered varieties Kompolti and Szegedi-9 were grown on the entire cultivation area in Hungary and superseded the older varieties. Since that time, total fiber yields in the fiber-processing industry have steadily increased. In the 1980s, the average fiber yield of this industry exceeded 24%, making it the highest in Europe. This high average is undoubtedly a direct result of breeding for higher fiber yields. The results even caught the attention of neighboring countries, where Kompolti was added to the national lists of registered varieties, first in the former Czechoslovakia and later in Bulgaria. The effectiveness of the selection was promoted through the methodical experiments conducted by István Jakobey.

The breeding of Kompolti hemp for fiber yield presently has the longest series of successive selection data because, since 1952, selection has been carried out on both parent lines (i.e., male and female).

The results from selection for increased fiber content show that the fiber yield of the elite plants in a widely spaced crop increased from 13.5% (1953) to 37% (1988). During the same period, the technical fiber content of densely grown descendant crops or varieties increased from 19.3% to 37.6%. Fiber yield of the industrial hemp stalk simultaneously increased from 18.0% to 24.2% (see Figure 12).

The increase in fiber content by selection is possible, because the genetic reserves of this variety for fiber content are very large; this can be further improved by preservation breeding.

Today, although it has had the same name for forty years, the Kompolti variety can be regarded as a new variety.

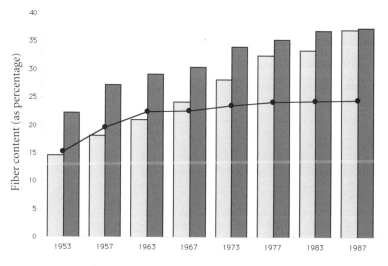

Figure 12: The influence of breeding on fiber content
of the variety Kompolti

4.3.2 The Breeding of a Unisexual Hemp Hybrid

A lack of mechanization in seed production combined with minimal financial interest have often led to a shortage of seeds in Hungary. In the 1960s and 1970s, seeds had to be imported from other countries on numerous occasions. These imports severely disrupted the hemp industry because yield capacity and fiber content of the imported varieties (e.g., Turkish varieties and French monoecious varieties) were much smaller than those of domestic varieties.

By then, the variety UNIKO-B had already been on the market for a few years. This variety is actually the F1 or F2 generation of a cross of a dioecious variety with a monoecious variety, and uses the unisexual trait to its advantage. This process was first carried out in Kompolt, Hungary. It is common knowledge that the F1 generation of dioecious × monoecious hybrids mainly consists of female and monoecious plants, i.e., seed-bearing plants. The ratio can be as high as 90–95%. The seed yield of such crops is 70–90% greater than that of dioecious varieties. Apparently, the idea was to use the large seed yields of hybrids to alleviate the threat of a seed shortage. The UNIKO-B hybrid is a cross between Kompolti and Fibrimon 21.

The F1 generation consists of 70–85% female plants, 10–15% monoecious plants, and about 1–2% male plants. Their crossing is carried out on a total of five hectares (12 acres). F1 seeds are sown on the seed acreage (500 hectares, or 1,200 acres). The seed requirement, therefore, is four kilograms per hectare (3.5 pounds per acre). The resulting high seed yield surplus obtained a national average in some years of 70% in comparison to the Kompolti variety in the 1970s. These were the industrial seeds that the fiber industry received for sowing; that is, the industry uses F2 seeds. It turned out that the UNIKO-B hybrid has a larger stalk yield in the F2 generation than does Kompolti hemp, and its average fiber content is only 1–2% less. Finally, UNIKO-B's total economic value exceeds that of Kompolti, thanks to its high seed yield. For a while, 80–90% of all hemp acreage was grown with the UNIKO-B variety. Figure 13 illustrates production and use of the UNIKO-B variety.

Histological and genetic constraints limit the increase of fiber content. Stalk yield can be significantly increased by simple selection, but greater possibilities are offered by heterosis (hybrid) breeding.

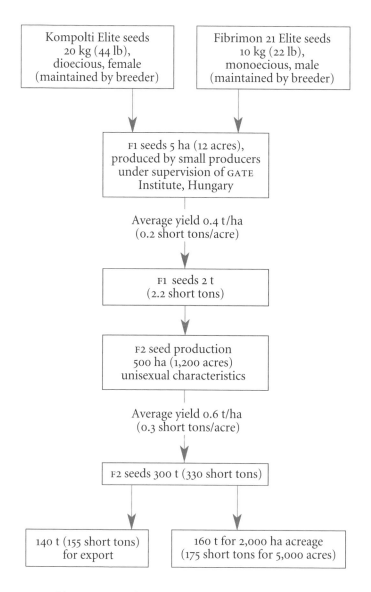

Figure 13: Production and use of UNIKO-B seeds

4.3.3 The Breeding of Hybrid (Heterosis) Hemp

In the 1970s—after the discovery of unisexual hemp—the heterosis effect (also known as *hybrid vigor*) was realized in a two-way cross of three hemp varieties.

In Hungary the existence of the heterosis effect for hemp was already known in the 1960s. In the F1 generation, the hybrid derived from crossbreeding Kompolti and a Chinese variety proved to have an increased stalk yield of 12–15% in comparison to the Kompolti variety. This method could not be implemented because it was too hand-labor-intensive to emasculate the male parent. Thus, the production of unisexual Chinese female varieties was necessary, and this could only be accomplished by crossbreeding Chinese monoecious varieties with Chinese dioecious varieties.

A Chinese monoecious variety had to be bred first, which was accomplished by the mid-1970s. The heterosis effect proved that there was no other obstacle to producing unisexual hemp similar to the initial hybrid hemp, by means of a two-way cross between three different varieties. The improvement in stalk yield with TC (three-way-cross) hybrids was almost 10%. In 1982, these hybrids were included on the Hungarian list of registered varieties as Kompolti Hybrids TC, and they are now grown on large plots in Hungary because the female single cross (Chinese, unisexual) is a high-seed-bearing variety; seed production is thus profitable and reliable.

Figure 14 outlines the schematic for development and usage. Where a THC content of 0.3% or less is not legally imposed, growing such a hybrid is clearly favorable, because it has the highest stalk yield of any European variety and its fiber content is equally good. Thanks to their excellent seed yield, the cultivation of female single cross hybrids is profitable and reliable (1–1.2 tons per hectare, or 900–1,100 pounds per acre).

4.3.4 The Breeding of Monoecious Hemp

Monoecious varieties of hemp have never been cultivated in Hungary, or in the southern sections of Russia, Romania, former Yugoslavia, or Bulgaria. After many years of testing in Kompolt, Hungary, it was apparent that stalk yield and fiber yield per acre or hectare of the

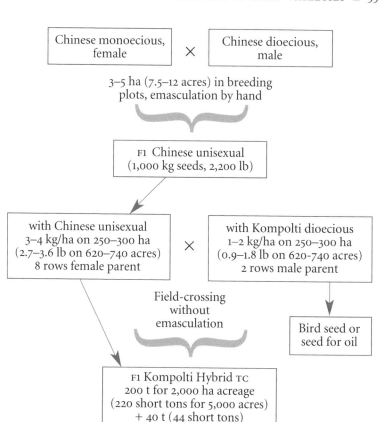

Figure 14: Schematic for the development and usage
of the variety Kompolti Hybrid TC

monoecious (Ukrainian and French) varieties remain lower than those of the dioecious varieties. One reason is the fact that monoecious hemp is capable of self-pollination, which results in a certain inbreeding depression, whereas dioecious hemp is exclusively cross-pollinated. Tables 3 and 4 show the stalk yield, fiber content, and fiber yield of monoecious and dioecious hemp. Table 4 shows that the relative yield, when monoecious and dioecious varieties are compared, has not changed over a period of fifteen years. The superiority of dioecious hemp can be interpreted as a biological axiom.

Table 3: Yield quantity of different varieties of hemp

Variety	Origin	Stalk Yield metric t/ha (lb/acre)	Fiber Content (%)	Fiber Yield metric t/ha (lb/acre)
Kompolti (dioecious)	Hungary	11.6 (10,300)	27.4	3.26 (2,910)
UNIKO-B (dioecious)	Hungary	12.3 (11,000)	25.1	3.10 (2,760)
JUSO-9 (monoecious)	Ukraine	8.1 (7,200)	25.1	2.03 (1,810)
Fibrimon 56 (monoecious)	France	9.8 (8,700)	21.4	2.09 (1,860)
Fibrimon 24 (monoecious)	France	8.9 (7,900)	23.5	2.08 (1,860)
LSD$_{5\%}$*		± 0.8	± 1.2	± 0.25
Test Trials	–	4	–	–

Table 4: Yield results of monoecious and dioecious varieties from Hungarian test crops (Kompolt)

Variety	Origin	Stalk metric t/ha (lb/acre)	Relative Yield (%)	Fiber Content (%)
1979				
Dioecious	Hungary	12.1 (10,800)	100	–
Monoecious	Ukraine	7.8 (7,000)	64.8	–
LSD$_{5\%}$*	–	± 0.75	± 0.64	–
1984				
Dioecious	Hungary	7.7 (6,870)	100	30.1
Monoecious	France	7.6 (6,780)	99.4	27.7
LSD$_{5\%}$*	---	± 0.60	± 0.80	–
1994				
Dioecious	Hungary	10.4 (9,280)	100	35.8
Monoecious	France	9.3 (8,300)	89.6	28.2
LSD$_{5\%}$*	–	± 1.55	± 1.49	–

*LSD$_{5\%}$ = least significant difference at p=0.05

In addition, monoecious hemp can never reach the standards of dioecious hemp by means of selection. First and foremost, its cultivation is only feasible in countries like France and the former Soviet Union, where hemp is cultivated for dual usage (i.e., cultivated for fiber and seeds). When dioecious hemp is grown for seed and fiber, the male plant stalks lignify and break before maturation of the female plants. This does not happen with monoecious hemp because every plant has female characteristics and they mature at the same time. In any event, this simultaneous maturation has no effect on the harvested fiber hemp stalks, which are still green.

Although monoecious hemp is not cultivated in Hungary, it is still bred because, as was previously stated, it yields the male parent line of the unisexual UNIKO-B variety. Its breeding evolves in two parallel directions. The first is the stabilization of monoecious hemp as an artificial construct that cannot exist without the aid of humans. In the course of evolution, hemp evolved as a dioecious form over thousands of years. Hence, this is the naturally occurring form of hemp. Although monoecious (intersex) forms occur at a rate of 0.1–0.2% in every dioecious population, this is recessive in heredity and will rapidly be blotted out by the dominant dioecious variety.

The Sengbusch Classification System is regarded as the definitive guide for selection of true monoecious types. According to this system, following selection, only second- or third-degree monoecious intersex types remain in the population. The predominantly male (first-degree) and predominantly female (fourth- and fifth-degree) types are removed before flowering (see Figures 15A and 15B).

According to the Sengbusch Classification System, the following monoecious forms exist, all with female traits:

- the first-degree monoecious type has 80-90% male flowers;

- the second-degree monoecious type has 60-70% male flowers;

- the third-degree monoecious type has roughly 40-50% male flowers (the second-degree and third-degree monoecious types, according to Sengbusch, are termed ideal monoecious types and are the basis for monoecious cultivation);

Figure 15A: Intersexes of monoecious hemp according to the Sengbusch Classification System: pure males, first-degree, second-degree, third-degree, fourth-degree, and fifth-degree monoecious types *(left–right)*

Figure 15B: First-degree, second-degree, third-degree, fourth-degree, fifth-degree monoecious types *(left–right)*

■ the fourth-degree monoecious type contains 10–30% male flowers (only on the primary axis, on a few branches, and on one or two secondary shoots);

■ the fifth-degree monoecious type has fewer than 10% male flowers (and can easily be mistaken for a female type, especially if the few male flowers have already fallen off).

This method of sex selection has been universally adopted by breeders of monoecious hemp. A monoecious rate of 99% can be attained. However, dioecious male plants occur at ever-increasing rates during subsequent generations, even if in the beginning they were not present in the crop population. Therefore, it is not possible in principle or in practice to produce a genetically pure and stable monoecious hemp as claimed by Professor Hoffmann in the 1950s.

The other goal of breeding is increased fiber content. The Bredemann method cannot be used to satisfy this goal in the case of breeding monoecious hemp; only selection of female plants is possible. Compared to selection for high fiber content in dioecious hemp, selection in monoecious hemp is only 50% as effective or less.

In Hungary, a later variant of the Fibrimon 21 monoecious variety of German-French origin is bred, or rather maintained. This is the male line of UNIKO-B. A Chinese monoecious variety is used to breed hybrid hemp.

4.3.5 The Breeding of Hemp With a Low THC Content

Delta-9 THC (tetrahydrocannabinol), which is classified as an intoxicant, can be found in lesser or greater quantities in every variety of hemp, including cultivated varieties. Since the THC content in cultivated varieties was restricted to 0.3% in most countries, particularly in European Community countries, it has become important to keep the THC content in older varieties below this limit and to not allow new varieties to reach the limit.

Within a particular variety, the THC content is largely dependent on environmental conditions. The same variety can produce much more THC at 46 degrees north latitude than in the cooler, rainier summer weather common in northern Europe (between 53 and 55 degrees north latitude). At elevations of no more than 200–250 meters (600–800 feet) above sea level, the plants' THC content is higher than in the same variety cultivated at an altitude of 500–600 meters (1,600–2,000 feet) above sea level.

With regard to THC content, there are also differences between the individual geographical races. With no breeding, the THC content of plants in the southern region is the lowest. Plants from the central Russian geographical race have a medium THC content, while plants from the Asiatic geographical race (not the Indian landrace cultivar!) have the highest concentration of THC. These last varieties are still classified as industrial hemp.

The breeding of low-THC hemp can be successfully carried out by rapid screening (thin-layer chromatography) of several hundred plants selected for specific purposes (fiber content, stalk yield, etc.). Plants that show no color reaction with this testing are further tested using gas chromatographic analysis. Only those plants not showing a reaction

Table 5: Delta-9 THC and CBD content in different
varieties of hemp and their THC/CBD ratio

Variety	Delta-9 THC (%)	CBD (%)	THC/CBD
Kompolti	0.15	1.39	0.11
JUSO 11	0.12	0.96	0.12
Felina 34	0.15	1.39	0.11
Futura	0.15	1.21	0.12
Ferimon	0.17	1.16	0.15
Fibrimon 56	0.21	1.07	0.19
UNIKO-B	0.22	1.21	0.18
Fedrina 74	0.25	1.67	0.15
Fedora 19	0.26	1.41	0.18
Bialobrzeskie	0.26	0.59	0.45
Kompolti Hybrid TC	0.55	0.79	0.71
Lovrin 110	0.66	1.30	0.51
Secuieni 1	0.75	1.14	0.66

De Meijer et al. 1992

with either of the analyses, the so-called minus variants, will be sown the following year. Since the bracts contain the most THC, chemical testing is conducted on these sections containing the sticky resin (Kompolt method). Fiber hemp has an easily identifiable characteristic that distinguishes it from marijuana grown for THC. Its nonpsychoactive CBD (Cannabidiol) content is significant and the THC/CBD ratio is less than 1. Table 5 outlines the delta-9 THC and CBD content as well as the ratio of THC to CBD in different varieties of hemp.

In the 1970s, France was the first country in Europe to begin cultivating hemp with a low THC content. The former Soviet Union (Ukraine-Gluckov) then followed suit. By the beginning of the 1980s, selection had begun in Hungary and Poland. Today, France grows varieties of hemp that are almost THC-free (0.05%), although the production of hemp that contains absolutely no THC is biologically inconceivable because THC plays an important, though not well understood, role in plant development.

4.4 Breeding Varieties and Their Evaluation

In the following discussion, we have assumed that two large ecological form groups and their transition forms exist in Europe (see Figure 16). Hungary, Romania, Serbia, Turkey, and parts of the former Soviet Union exclusively grow dioecious and unisexual varieties belonging to the southern ecological form group.

The selection of the variety plays a critical role in hemp cultivation (to a greater extent than with other cultivated plants) because there are considerable differences between varieties for stalk yield (early- or late-maturing), fiber content, seed yield, etc. A poorly selected variety can result in a fiber yield reduction of 30–40% per acre or hectare, whereas a well-chosen variety can result in an equal increase in profitability. First-time hemp growers will undoubtedly have difficulties selecting the right variety.

The southern varieties have the highest stalk yield (10–12 tons per hectare, or 4.5–5.4 short tons per acre) and are the latest to reach technical maturity. Their seeds do not mature above 50 degrees north latitude. They have a high fiber content and fine fibers. The vast majority of plants from this group are dioecious and represent typical

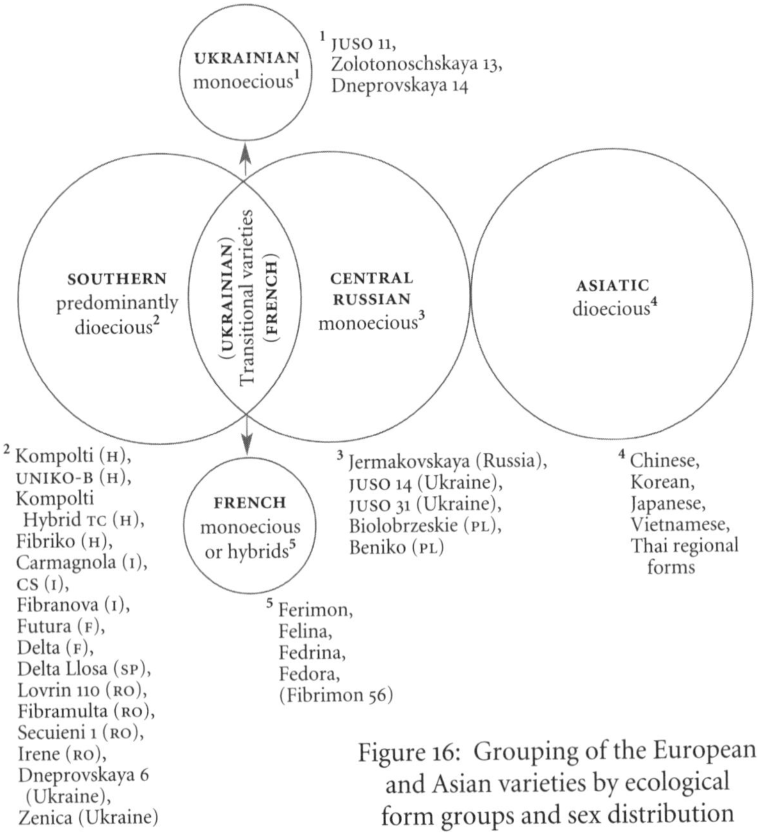
UKRAINIAN
monoecious[1]
[1] JUSO 11,
Zolotonoschskaya 13,
Dneprovskaya 14
SOUTHERN
predominantly
dioecious[2]
(UKRAINIAN)
Transitional varieties
(FRENCH)
CENTRAL
RUSSIAN
monoecious[3]
ASIATIC
dioecious[4]
[2] Kompolti (H),
UNIKO-B (H),
Kompolti
Hybrid TC (H),
Fibriko (H),
Carmagnola (I),
CS (I),
Fibranova (I),
Futura (F),
Delta (F),
Delta Llosa (SP),
Lovrin 110 (RO),
Fibramulta (RO),
Secuieni 1 (RO),
Irene (RO),
Dneprovskaya 6
(Ukraine),
Zenica (Ukraine)
FRENCH
monoecious
or hybrids[5]
[5] Ferimon,
Felina,
Fedrina,
Fedora,
(Fibrimon 56)
[3] Jermakovskaya (Russia),
JUSO 14 (Ukraine),
JUSO 31 (Ukraine),
Biolobrzeskie (PL),
Beniko (PL)
[4] Chinese,
Korean,
Japanese,
Vietnamese,
Thai regional
forms

Figure 16: Grouping of the European and Asian varieties by ecological form groups and sex distribution

fiber- or cellulose-rich varieties of hemp. The vegetative period is calculated from the beginning to the middle of flowering (50% flowering of male flowers) because harvesting must then be completed. Under no circumstances should the seeds mature, or even form. Southern varieties require 105–115 days from germination to the middle of male flowering. At 45 degrees north latitude, flowering occurs from mid- to late June. Above 50 degrees north latitude, flowering usually occurs from mid- to late August. The seed yield is mediocre to below average because half of the crop (the males) produce no seeds. The seeds mature in 140–170 days. These varieties can also be successfully harvested in the north because their flowering is delayed by the long summer days, which increases the stalk yield as with early-maturing varieties. Varieties in this region include the Hungarian, Romanian, south Russian, Italian, Spanish, and Turkish bred and non-bred varieties, as well as one French and two Russian monoecious varieties.

The other large group consists of the central Russian types. These varieties have a stalk yield of 5–6 tons per hectare (2.2–2.7 short tons per acre), low fiber content, and coarse fibers.

The central Russian varieties require 70–80 days to male flowering and 105–110 days to reach seed maturity, because natural selection and the short vegetative period have altered their development. This group includes certain Ukrainian varieties, as well as French and Polish varieties. The former All-Union Research Institute for Fiber Plants was located in Gluckov in the Ukraine. In Russia, southern varieties of hemp are presently bred only in the Research Institute in Krasnodar. The majority of varieties of this group are monoecious.

4.4.1 Transition Varieties

The varieties belonging to both the southern and the central Russian type were first crossbred with one another in the former Soviet Union, then in Germany and France. From these crosses, the so-called "southern varieties that also mature in the north" were developed. This claim should not be taken literally, because the vegetative period and other value-determining characteristics of these transition varieties between the central Russian and southern regions play an important role. These varieties are almost exclusively monoecious or hybrid populations. Many are cultivated for dual usage. The vegetative period is 85–95 days to male flowering and 115–130 days to seed maturity. Potential stalk yield (7–9 tons per hectare, or 3.1–4.0 short tons per acre) and fiber content are only mediocre, while seed yield is good to very good. Varieties included in this classification are all of the French varieties (except Futura) and the Ukrainian varieties JUSO 11 and Zolotonskaya 13. The THC content of Ukrainian and French varieties is very low.

4.4.2 French Varieties

Only a small proportion of French varieties are monoecious. The majority belong to the so-called "hybrid population." They are all included in the European Union Registered List of Varieties, and thus have a leading position in western Europe. Because of their vegetative period and stalk yield, they represent a transition between the southern and central Russian types, but are more closely related to the southern types. Their stalk yield does not match that of the southern varieties, their fiber content varies but is generally mediocre, their seed yield is good, and their THC content is very low. The earliest-maturing variety is the monoecious Ferimon 12; the latest-maturing is the monoecious Futura. Without exception, the French varieties are grown in their native country for the paper industry, and a portion are planted for dual usage (stalk and seeds are harvested from the same crop).

Marketing and sales of the French varieties are exclusively conducted by the Fédération Nationale des Producteurs de Chanvre (FNPC) 20, Rue Ligneul, 72003 Le Mans, France.

The French "hybrid populations" are basically pseudohybrid "unisexual" varieties that were cultivated because of the difficulties of sex stabilization with monoecious hemp. Maintenance of these varieties is much easier than with monoecious varieties, and their reproduction is much more rapid and more reliable. In comparison to the Hungarian monoecious varieties (UNIKO-B), the end product includes 50% female plants and 50% monoecious plants, and the male proportion is insignificant. The production pattern is as follows:

Parental generation:	Dioecious . . . × Monoecious . . .
FI generation:	Unisexual . . . × Monoecious . . .
F2 generation:	Seeds for sale on the market

According to the Hungarian method, the unisexual FI with open-pollinated male plants produces the F2 generation, which is almost completely dioecious. Other methods are used in the production of a hybrid population, whereby the unisexual FI is crossed with monoecious hemp, which results in seed for sale on the market. Such hybrid populations include the French varieties Fedora 19, Felina 34, and Fedrina 74. Today the only true monoecious varieties are Ferimon and Futura, which proves our prior claim that it is very difficult, if not impossible, for monoecious hemp to persist without the reoccurrence of male plants.

Monoecious Varieties

■ *Ferimon (F 12)*

Derived from Fibrimon 21, Ferimon is the earliest-maturing French variety. Its stalk and fiber yield are both mediocre, while its fiber content is good. Its seed yield is very good, but still less than that of Fedora 19. Cultivation is recommended in light soil where seeds mature through mid-September. This variety was bred by J. P. Mathieu.

■ *Futura (F 77)*

This is the latest-maturing French variety. It is derived from Fedrina 74, and has the highest potential stalk yield of the French varieties and a mediocre fiber content. Futura is not suitable for market seed production. It is harvested green (as fiber hemp). Rich soil and a region with plenty of rainfall during the vegetative period are recommended in order to achieve full yield potential. Futura was bred by J. P. Mathieu.

Hybrid Populations

■ *Fedora 19 (F 19)*

Derived from the monoecious Fibrimon 21 variety and a dioecious central Russian variety, this early-maturing variety has a mediocre stalk yield and fiber content, but a very high potential seed yield, making it very well-suited for seed production. Fedora 19 was bred by M. Arnoux and J. P. Mathieu.

■ *Felina 34 (F 34)*

An early- to mid-maturing variety equally suited for fiber harvesting and seed production, Felina 34 is derived from Fibrimon 24 and a German dioecious variety. It has a good potential stalk yield and a good fiber content, and its potential seed yield is even better than that of Fedora. In France, the seeds mature by the end of September. This is the most prevalent French variety, and it was bred by M. Arnoux and J. P. Mathieu.

■ *Fedrina 74 (F 74)*

A late-maturing variety that is appropriate for green harvesting, Fedrina 74 is derived from Fibrimon 24 and a Hungarian dioecious variety. It has a good stalk/fiber yield, but its fiber quality is only mediocre. This variety is recommended for medium to fertile soils and regions where the weather is poor (i.e., very rainy) by the end of August. It was bred by M. Arnoux and J. P. Mathieu.

4.4.3 Hungarian Varieties

Without exception, the Hungarian varieties all belong to the southern type. They have a long vegetative period, are dioecious, and are open-pollinated or hybrids.

■ *Kompolti*

Derived from Fleischmann Hemp, Kompolti is the oldest, best-known open-pollinated dioecious variety in Europe. It was nationally registered in Hungary in 1954. Since then, its fiber content has been increased by 250% and it now contains 35-38% technical fibers. Its potential stalk yield is very good, with a maximum yield of 11–12 tons per hectare (4.9–5.4 short tons per acre); its potential seed yield is mediocre. Thanks to its high fiber content, Kompolti is particularly well suited for the industrial paper industry. It is grown only as fiber hemp and cut green at the beginning of August. As fiber hemp, Kompolti's vegetative period is 110–115 days. Its THC content is very low (0.1–0.15%), and it was bred by I. Bócsa and J. Schmidt.

■ *UNIKO-B*

This variety is a hybrid from Kompolti and the German monoecious Fibrimon 21; the latter is unisexual in the F1 generation (the population consists of female plants, monoecious plants with female appearance, and a small proportion of male plants). Using the high potential seed yield of F1, F2 is introduced to the market as fiber hemp. Its vegetative period as fiber hemp is 105–110 days. The stalk yield is very good (a maximum of 11–12 tons per hectare, or 4.9–5.4 short tons per acre); the fiber content is good but less than that of Kompolti. Industrial F2 hemp is not suited for seed cultivation because it cannot be reproduced. The THC content is low (0.2–0.25%). It was bred by I. Bócsa.

■ *Kompolti Hybrid TC*

This variety is a three-way cross hybrid with the unisexual F1 generation of Chinese monoecious × Chinese dioecious varieties. Since these have no males, the Kompolti hemp is generally crossbred without emasculation (TC = three-way cross). Its vegetative period is the longest with 120–125 days, and its stalk yield is greater than that of Kompolti (a maximum of 12–13 tons per hectare, 11,000–12,000 pounds per acre) while its fiber yield is lower (30–33%). The THC content is 0.4–0.5%, so it can only be grown in Hungary or other countries where the restriction of the European Union to 0.3% is not in effect. Due to its hybrid character, general reproduction is not possible. This variety was bred by I. Bócsa, S. Bata, E. Horkay, and I. Györki.

■ *Fibriko*

Fibriko is a three-way cross (TC) with light-green leaves. The single-cross parents correspond to those of Kompolti Hybrid TC; its third variety, however, is Kompolti Sárgaszárú (which has a light-colored stalk). Fibriko's vegetative period is 100–105 days. Its stalk yield is good, its fiber content is very good (33–35%), and its THC content is 0.3–0.4%. Of all Hungarian varieties, Fibriko has the finest and strongest fibers. Since it is a hybrid, it cannot be further reproduced. This variety was bred by E. Horkay, S. Bata, and I. Bócsa.

■ *Kompolti Sárgaszárú (light-colored stalk)*

This variety is derived from multiple backcrossing of light-colored dwarf hemp from Hoffmann and green Kompolti. The leaves and the stalk lose their chlorophyll content before flowering and become butter- to lemon-yellow. This is very favorable for fiber and paper production. Its vegetative period is 95–100 days. The stalk yield of Kompolti Sárgaszárú is roughly 15% less than Kompolti's but its fiber content is equal to that of Kompolti. This is an open-pollinated variety, from which the green plants (fewer than 1%) are removed during seed production. The THC content is 0.2–0.3%. This variety is not yet available on the market, and was bred by I. Bócsa and S. Bata.

All Hungarian germplasm is preserved at the Fleischmann Rudolf Agricultural Research Institute at GATE, H-3356 Kompolt, Hungary. The marketing firm is Fibro-Seed GmbH, H-3356 Kompolt, Hungary.

4.4.4 Romanian Varieties

The Romanian varieties belong exclusively to the southern type. The Romanian list of registered varieties includes two dioecious and two monoecious varieties of hemp.

■ *Fibramulta 151*

This is the oldest Romanian variety (1965). Fibramulta 151 is a dioecious variety derived from the cross of a Romanian strain with the German Fibridia variety. Its vegetative period is 115–117 days, its potential stalk yield is good (a maximum of 8–10 tons per hectare, or 3.6–4.5 short tons per acre), and its fiber content is 26%. This variety's potential seed yield is good for a dioecious variety (a maximum of 800 kilograms per hectare, or 1,750 pounds per acre), and its fiber quality is equally good. The THC content is unknown. It was bred by N. Ceapoiu and E. Itoafa.

■ *Lovrin 110*

This is a dioecious variety derived from a Bulgarian population. Its vegetative period as fiber hemp is 110–115 days. Lowrin 110's stalk yield is good (a maximum of 9–11 tons per hectare, or 4.0–4.9 short tons per acre), its fiber content is mediocre at 27–30%, and its fiber quality is good. The THC content is unknown. This variety was bred by P. Parashivoiu.

Both of the above varieties are maintained by the Agricultural Breeding Facility, Lovrin (Komitat Timis).

■ *Secuieni 1*

This is a monoecious southern variety derived from a cross of the dioecious Dneprovskaya 4 and Fibrimon, and then selected for a monoecious variety. The population consists of 50–60% female, 30–40% monoecious, and 5–10% dioecious male plants. This variety's vegetative period is 105–110 days. Its potential stalk yield is good (a maximum of 8–9 tons per hectare, or 3.6–4.0 short tons per acre), its fiber content is very good (30–33%), and its potential seed yield is excellent (1,000–1,200 kilograms per hectare, or 900–1,100 pounds per acre). The THC content is unknown. The germplasm is maintained by the Agricultural Research Facility at Secuieni (Komitat Neamt). The variety was bred by N. Gáucá.

■ *Irene*

This new monoecious southern variety was registered in Romania in 1994. Irene blossoms later than Secuieni 1 (by about 10 days). Its maximum potential stalk yield is 11–12 tons per hectare (4.9–5.4 short tons per acre), its fiber content is 26–27%, and its potential seed yield is a maximum of 1.4 tons per hectare (1,250 pounds per acre). The THC content is unknown. This variety is maintained by the Agricultural Research Facility at Secuieni, and was bred by N. Gáucá.

4.4.5 Ukrainian Varieties

■ *Ermakovkaya Mestnaya*

A central Russian early-maturing variety that is primarily cultivated in West Siberia, Ermakovkaya Mestnaya can considered a national variety.

■ *JUSO 31*

This is an early-maturing variety with a vegetative period seven days less than that of JUSO 14. Its maximum stalk yield is 9 tons per hectare (4.0 short tons per acre), its fiber content is 25–26%, and its THC content is very low (0.04–0.07%). This variety was nationally registered in 1987, and was bred at the Research Institute for Fiber Plants in Gluckov by W. G. Virovez, G. I. Sentshenko, I. I. Seherban, and L. M. Gorshkova.

■ *JUSO 14*

JUSO 14 is an early-maturing monoecious variety from the central Russian region. Its vegetative period is 107–115 days to seed maturity and 90 days to technical maturity. Its stalk yield is a possible maximum of 8–9 tons per hectare (3.6–4.0 short tons per acre), it has a fiber content of 27–28%, and its potential seed yield is 1.1 tons per hectare (1,000 pounds per acre). The THC content is 0.16%. This variety was cultivated at the Research Institute for Fiber Plants in Gluckov, and bred by G. I. Sentshenko, W. G. Virovez, I. I. Seherban, and L. M. Gorshkova.

■ *Zolotonskaya JUSO 11*

This is a monoecious hybrid variety more closely associated with the southern type. Its vegetative period to maturation is 135–140 days. Its potential stalk yield is a maximum of 9 tons per hectare (4.0 short tons per acre), its fiber content is 27%, and its seed yield is 0.5 tons per hectare (450 pounds per acre). This variety, nationally registered in 1984, was developed in the Research Facility at Zolotonska and bred by G. I. Sentshenko, W. G. Virovez, M. M. Orlov, and W. A. Dishlevij.

■ *Zolotonskaya 13*

This variety is a monoecious hybrid more closely associated with the southern region. Its vegetative period to maturation is 140–145 days. This variety has a good potential stalk yield and a good fiber content (27.4%). The THC content is less than 0.2%. Zolotonskaya 13 was bred in the Research Facility at Zolotonska by N. M. Orlov and L. G. Orlova and nationally registered in 1986.

■ *Dneprovskaya 6*

This is a monoecious variety closely related to the southern type. Seed maturation occurs after 125–150 days; technical maturity after 105–120 days. Dneprovskaya 6 has a high fiber content (31%). Its seed yield is 0.6–0.9 tons per hectare (550–800 pounds per acre) and its THC content is 0.15–0.20%. The breeding of this variety was a cooperative effort by the Research Institute for Fiber Industry in Gluckov and the Research Facility at Sinelnikov. Dneprovskaya 6 was added to the national list of registered varieties in 1980, and was bred by R. J. Kaplunova, G. I. Sentshenko, W. G. Virovez, and L. M. Gorshkova.

■ *Zeica*

This is a southern dioecious variety with a vegetative period 10–12 days longer than that of Dneprovskaya 6. Compared to that variety, its stalk yield is larger, its fiber content is equal, and its fiber quality is even better. Zeica was nationally registered in 1990. It has a low seed yield, and its THC content is less than 0.2%. It was developed by W. A. Nevinnich, P. W. Nimtshenko, and G. W. Suchorada.

4.4.6 Polish Varieties

■ *Bialobrzeskie*

This monoecious variety was derived from a cross between Kompolti and a monoecious variety with which it was back-crossed several times. Nationally registered in 1986, Bialobrzeskie has a very good potential seed yield (800–1,000 kilograms per hectare, or 700–900 pounds per acre). Its potential stalk yield is a maximum of 10–12 tons per hectare (4.5–5.4 short tons per acre). The fiber content is mediocre (27–28%), yet the fiber quality is good. The THC content is very low. Developed by P. Obara, A. Strzelecki, J. Mikolajczyk, and H. Mackiewicz, this variety is maintained at the Institute for Natural Fibers (Poznan).

■ *Beniko*

Nationally registered in 1985, this variety has a higher potential seed yield than Bialobrzeskie and is extremely fiber-rich (37%). Its fiber yield can reach 3 tons per hectare (1.3 short tons per acre), and its THC content is very low. Beniko is maintained at the Institute for Natural Fibers (Poznan), and was bred by B. Jaranovska and J. Kozak.

4.4.7 Asiatic Varieties

The Asiatic region includes national varieties from China, Japan, Korea, Vietnam, and Thailand. These countries do not produce cultivated varieties of hemp; they grow landraces. These varieties branch out better in a widespread plot. They achieve only lesser heights and their vegetative period is more than 170 days. Many of these varieties do not even reach maturity in central and southeastern Europe. Their THC content is generally higher than that of varieties from other regions (0.6–1.0% THC), but the plants cannot be used to produce hashish. The plants have adapted to the shorter days at 35 degrees north latitude and, due to the high temperatures, mature within the vegetative period. Their fiber content is low and their seed yield is good. The Asiatic varieties have minimal economic significance in Europe.

Chapter 5

■ ■ ■ ■ ■ ■ ■

Hemp Cultivation

5.1 Climate and Soil Requirements

5.1.1 Temperature

The varieties of hemp cultivated in central and southeastern Europe belong to the southern (Mediterranean) ecological form group. The total heat quantity received by a plant in a growing period is important for its development. This total heat quantity is measured in growing degree days (GDD), and can be calculated as the sum of average temperatures of all days in the entire growing period. Fiber-hemp plants require a total heat quantity over the growing period of 1,900–2,000 GDD_C (3,400–3,600 GDD_F) from germination to technical maturity (110–115 days), and 2,700–3,000 GDD_C (4,900–5,400 GDD_F) until seedlings develop. Early-maturing varieties such as "Felina 34" require only 1,600–1,700 GDD_C (2,900–3,100 GDD_F) to reach technical maturity.

Hemp seeds begin to germinate when the ground temperature has increased to 1 or 2 C (34–36 F). If the ground temperature increases to 8–10 C (46–50 F), the seeds sprout in eight to twelve days. The color of the new leaves is yellowish to grayish green. At this developmental stage up to the fourth and fifth pair of leaves, the plants can survive frost down to a temperature of about –5 C (23 F), but they will stop growing, even if a period with rapidly warming temperatures follows. Thus, with Germany's weather conditions, it does not pay to sow too early (it is better to wait until the second half of April).

According to growth chamber tests conducted in Holland, plants that are grown for about forty days at a constant temperature of 19 C (66 F) reach the same height as plants grown for ninety days at a constant temperature of 10 C (50 F) (see Figure 17).

The test results demonstrate that the temperature range for optimal growth is between 19 and 25 C (66–77 F), and these results concur with estimates provided by experienced cultivators. If the average daily temperature reaches 10–15 C (50–59 F), the plants provide substantial ground cover, and sprouting weeds die off because they are completely overshadowed by the hemp foliage. If the average daily temperature reaches 16 C (61 F), the plants enter into the rapid growth stage. During this stage, the plants can reach a daily growth rate of 4–6 centimeters (1.5–2.5 inches) (see Figure 17).

Throughout Germany, there is a sufficient total heat quantity for both southern and transition varieties of hemp to reach technical maturity. However, due to inadequate daylight periods and temperatures, seeds of southern varieties of hemp do not mature. Seeds from the hybrid variety mature in the central Russian region if the necessary conditions are met during the vegetative period. Mid- to late-maturing varieties achieve seed maturation in southern and central Germany if they are grown below 53 degrees north latitude, while the early- and very-early-maturing varieties achieve seed maturation if grown below 55 degrees north latitude.

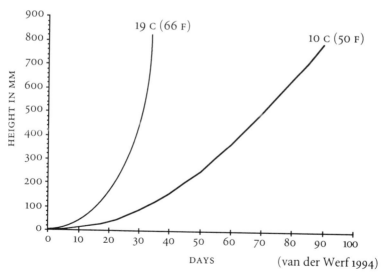

Figure 17: Growth of hemp in a growth chamber at temperatures of 10 and 19 C (50–66 F)

5.1.2 Light

The daylight conditions in Germany are suitable for cultivation of all industrial hemp varieties, including the southern, central Russian, and transition varieties. Two factors contribute to the light requirements: the total amount of sunlight during the vegetative period and the daily period of sunlight. Although overall there are not as many hours of sunlight in Germany as in southern Europe during the vegetation period, the amount is adequate even for southern varieties of hemp, because the amount of dry matter produced is determined not so much by the total amount of sunlight as by the daily period of sunlight. The total number of hours of sunlight during the vegetation period influences the harvest volume to a lesser extent than it does the fiber quality (primarily fiber strength).

During the vegetative period, the number of daylight hours is considerably greater in central and northern Germany than in central or southeastern Europe. As a result, the typical southern varieties of hemp cultivated in Germany have a larger stalk yield than do the central Russian or transition varieties. The photoperiodic reaction of hemp provides one explanation for this.

As was previously mentioned in Section 3.3.3, hemp is a short-day plant. This means that hemp enters its reproductive phase earlier when day length is short than it does when days are longer. In extreme cases, as was also noted in the Dutch tests, the southern varieties do not even flower by the end of August if they are subjected to twenty-four hours of sunlight. If growth and dry-matter accumulation continue any longer, the yield can be increased by 2–3 metric tons of dry matter per hectare (0.9–1.3 short tons of dry matter per acre). By this point, other varieties of hemp have long since stopped growing.

Table 6 outlines the number of days from germination to flowering, depending on day length. The results indicate that, with natural lighting, the difference in the period of time to flowering between the two varieties amounts to forty days. In this experiment, the central Russian variety still flowered at seventeen hours of exposure to light, whereas the southern variety did not flower under these light conditions. With sixteen hours of light, the southern variety began flowering after ninety-six days, whereas the central Russian variety had not yet

Table 6: The number of days from germination to male flowering, dependent on the hours of daylight

Daylight hours	Central Russian variety	Southern variety
Control (field test)	62	104
17	62	not flowering
16	62	96
15	57	89
14	41	52
12	37	42
10	34	39
8	34	41

Davidjan 1971

reacted. As the days grew shorter, both varieties took less time to reach the flowering stage, but the central Russian hemp still flowered earlier than did the southern variety. In addition, the shorter days caused the southern hemp to branch out more, whereas the central Russian variety did not branch out at all.

5.1.3 Water

Fiber hemp can produce 12–15 metric tons of dry matter per hectare (5.4–6.7 short tons per acre) from the entire fresh mass of leaves, stalks, and roots. According to studies, 500–700 millimeters (20–28 inches) of precipitation, or an adequate quantity of water, are necessary. In fact, fiber hemp requires at least 250–300 millimeters (10–14 inches) of precipitation during the vegetative period (soil water is accounted for in this calculation).

The water-absorption rate of fiber hemp is greatest during its fast-growth period. Hemp, with its extremely fast growth rate, demands a suitable temperature as well as an abundance of nutrients and water. The plant's specific water requirement is relatively large. Hemp requires 300–500 liters (80–130 gallons) of water for the production of 1 kilogram (2.2 pounds) of dry matter. The yield is predominantly influenced by the amount of rainfall during the months of June and July. Soil quality and specific soil preparation play an important role in satisfying hemp's substantial need for water.

One beneficial characteristic of hemp is the fact that its roots can adjust to the soil conditions. As long as the roots are not obstructed by the water table or by a hard layer of soil, the main root can reach a depth of 2–3 meters (6.5–10 feet). Thus, the roots can retrieve water from deeper soil layers. Once the plants have covered the soil, they utilize precipitation more efficiently than any other cultivated plant.

Within a field of hemp plants, warming occurs slowly; if there is little wind, the evaporation of precipitated water is minimal. Following heavy rainfall, the plant stock will be destroyed after only one or two days if water remains standing in puddles on the lower parts of the saturated field.

5.1.4 Soil

Fiber hemp cultivated for high yield and high quality has special requirements regarding soil quality. In other words, real "fiber hemp soils" are those that potentially yield 10 metric tons of dry matter per hectare (4.5 short tons of dry matter per acre).

Taxonomically,[1] the soil types with the highest yield are primarily the ones that developed on loess: very rich black soils (mollisols), degraded black soils, brown rendzina soils, and brown steppe soils. These soils have a favorable water balance, good water permeability, and an excellent nutrient-accumulation potential. The "brown" soils and the transitions to black soils are also suitable, provided that the soil is deep enough. "Pseudogley" soils (gray, mottled soils with reduced iron and other elements) are not suitable, because they are too wet and acidic. The ideal soil acidity for hemp is between pH values of 5.8 and 6.0. Many of these soil types originated on sandy loess. Podsol soils can be classified as third-rate soils, and are suitable if they did not originate from wind-deposited (eolian) sand and if the sand layer is not too deep. Sandy soils are mostly predominant in areas where rainfall amounts

---

[1] The soil descriptions are based on the European soil classification system; whenever possible a description or an equivalent to the USDA soil classification system has been given.

exceed 300 millimeters (12 inches) during the vegetative period from April to September, or in areas having a water table 80 centimeters (2.5 feet) or less below the surface. Using a scale based on yield per cultivated acreage in Germany, the following soil textural classes are suitable for the cultivation of hemp:

1S	Loamy sand (*D* Deluvial, *Loe* Loess, *Al* Alluvial, *V* degraded soils, *Vg* coarse degraded soils)
S1	(S/1S) Sandy loam (*D, Al, V*)
SL	(1S/sL) Loam (*D, Loe, Al, V, Vg*)
sL	Silt loam (*D, Loe, Al, V, Vg*)
L	Silt (*D, Loe, V, Vg*)
LC	Silty clay, silty clay loam (*D, Al, V, Vg*)

Figure 18 is a map of soil types in Germany with regard to their suitability for the cultivation of hemp. However, since a variety of other factors such as climate play a role in the cultivation of hemp, the categorization of areas by soil quality is not absolute.

Suitable soils for hemp cultivation are found in mountainous regions and low mountain ranges (different types of brown soils, and rendsolls, a variety of mollisols). However, the slope should not exceed 5%. If the slope is greater than 5%, precipitation runs off the field, especially since the water retention capacity of these soils is not very good. This could result in a temporary drought that causes damage to the crop. It is important that the fields face south. Only then can the stalks dry sufficiently on the stubble between the end of August and the beginning of September.

If the crop is to be dew-retted, altitude also plays an important role. The late-maturing varieties of hemp are not cultivated at altitudes higher than 200–300 meters (650–1,000 feet) above sea level, because temperatures are not adequate and problems concerning the drying process can occur. The early-maturing varieties can be cultivated at altitudes of up to 400 meters (1,300 feet) above sea level. At higher altitudes, technical maturation and the drying process are at risk, even for early-maturing varieties. The critical points to keep in mind are the 5% slope and an altitude of 400 meters (1,300 feet) above sea level.

CLASS I   CLASS II   CLASS III   OTHER SOIL TYPES

Figure 18: Map of soil types in Germany outlining I-, II-, and
III-rated soils suitable for the cultivation of hemp

It is important to note that these are only estimates. Therefore, it is advisable to grow test crops on as many different areas as possible and to take production samples in order to gather as much information as possible. In any event, it is better to cultivate hemp in the lowland rather than in the mountain regions.

It is especially important to be aware that the marshy soils in northern Germany are not at all suitable for the cultivation of textile hemp. Such marshy soils result in poor fiber quality and reduced tensile strength. It has not been determined whether these soils will produce fiber that can be used by the cellulose industry. Since such experiments have never been conducted, a variety of tests would need to be conducted to determine this.

Finally, pasture soils may be suitable for hemp cultivation; however, the water table in pastures is dependent upon the water level of the river. This means that hemp easily dies off in such areas because it can tolerate neither high water tables nor the stagnation of water. Hemp should be cultivated on pastures only if the land is protected by a dike and the crop is planted as far away from the dike as possible.

5.2 Nutrients

Soils are categorized into groups according to the amount of nutrients required by hemp and by the estimated yield per acreage. The soils are categorized as first-class soils (class I) or second- and third-class soils (class II). Class III soils have been included in class II together with second-class soils. A yield of 6–9.5 metric tons per hectare (2.7–4.2 short tons per acre) can be expected on class I soils and a yield of 4–7 metric tons per hectare (1.8–3.1 short tons per acre) on class II and III soils (see Table 7).

Fiber hemp's high requirement of nutrients is due to its high yield of dry matter. The crop also places high demands on the quality of the nutrients, which must be easily absorbed as well as available in large quantities, due to the relatively short vegetative period (about 110–115 days) of hemp.

During its rapid-growth period, hemp already requires considerable amounts of nutrients, and thereafter the requirement for nutrients even increase until the plants reach technical maturation, i.e., until full flowering of the male plants.

Table 7: Stalk yield of hemp by yield category for different soil classes

Yield Category	Stalk Yield		Class II and III Soils metric tons/ha (short tons/acre)
	Class I Soils metric tons/ha (short tons/acre)		
(1)	≤ 6.0	(≤ 2.7)	≤ 4.0 (≤ 1.8)
(2)	> 6.0–6.5	(> 2.7–2.9)	> 4.0–4.5 (> 1.8–2.0)
(3)	> 6.5–7.0	(> 2.9–3.1)	> 4.5–5.0 (> 2.0–2.2)
(4)	> 7.0–7.5	(> 3.1–3.3)	> 5.0–5.5 (> 2.2–2.5)
(5)	> 7.5–8.0	(> 3.3–3.6)	> 5.5–6.0 (> 2.5–2.7)
(6)	> 8.0–8.5	(> 3.6–3.8)	> 6.0–6.5 (> 2.7–2.9)

At the point of technical maturity, the male plants of dioecious varieties no longer take up nutrients, but the female plants continue to utilize nutrients until their seeds mature.

Numerous tests conducted in Hungary and other countries showed similar results for nutrient requirements.

According to available data and the opinion of various authors, the minimal and maximal quantities for the production of one metric ton of hemp stalks per hectare (0.5 short tons per acre) are: 15–20 kilograms (3–4 pounds) nitrogen, 4–5 kilograms (9–11 pounds) P_2O_5 (phosphate), and 15–20 kilograms (33–44 pounds) K_2O (potassium oxide).

However, chemical analyses can provide only a rough estimate of the required nutrients. Natural conditions and processes in the soil such as desorption, ion-exchange capacity, and washout also influence the nutrient requirements. A number of published references agree that the necessary amount of nutrients is considerably greater than the amount derived from analyses of dry hemp plants.

■ Nitrogen

Because of its vital role in growth, nitrogen is considered the most important nutrient for hemp. As soon as fiber hemp begins to grow, it requires large amounts of nitrogen. A yellowish grey-green color of the first true leaves on the seedling indicates a lack of nitrogen. It is typical for hemp to develop many leaves during the first half of the rapid-growth period. The stalk remains relatively small at this developmental stage. An adequate supply of readily available nitrogen must be provided to the plants throughout the entire vegetative period in order to achieve a high stalk yield. Excessive amounts of nitrogen, however, can be detrimental to the development of tissue and to the overall stalk yield. When the required amount of nitrogen has been exceeded, the stalk becomes larger, the bark section is thinner, and the fiber content and strength are reduced.

■ Phosphorus

Because of the increase in stalk yields and the development of fiber bundles, fiber hemp requires a steady supply of phosphorus from the point of germination to harvesting. Phosphorus not only plays a critical role for the total yield but is also important in forming the fiber cells. The need for phosphorus steadily increases until the plant reaches the flowering stage. Phosphorus contributes to the elasticity and tensile strength of the fiber cells or fiber bundles. Additionally, the plants need phosphorus in order to effectively utilize nitrogen. Table 8 summarizes the available amounts of phosphorus (P_2O_5) according to soil quality.

Table 8: Availability of soluble P_2O_5 depending on soil quality

Soil Quality	Aluminum-soluble P_2O_5 (ppm)				
	Very Poor	Poor	Adequate	Good	Very Good
Class i	< 50	< 90	< 150	< 250	< 450
Class ii	< 40	< 80	< 130	< 200	< 400
Class iii	< 30	< 60	< 100	< 150	< 350

■ Potassium

Fiber hemp's potassium requirement is also substantial. From germination to harvesting, potassium intake increases continually, and it is highest during the development of fibers in June and July. Fiber quality is affected to a greater extent by potassium than by phosphorus. Table 9 illustrates the amount of soluble potassium oxide in soils of varying qualities.

Table 9: Availability of soluble K_2O depending on soil quality

Soil Quality	Aluminum-soluble K_2O (ppm)				
	Very Poor	Poor	Adequate	Good	Very Good
Class I	< 150	< 250	< 380	< 500	< 700
Class II	< 100	< 160	< 240	< 350	< 550
Class III	< 80	< 130	< 200	< 300	< 500

■ Other Nutrients

Specialized literature contains details about the calcium (CaO) and magnesium (MgO) requirements of fiber hemp. But, in practice, these nutrients are not given specifically to one crop, but are applied as required by the particular crop rotation.

The role of trace elements in the cultivation of hemp has never been clarified. Leaf fertilization with trace elements has recently become customary practice, but the results are indefinite. Preliminary investigations suggest positive results from leaf fertilization during dry periods, but definite data concerning the yield and fiber quality of the resulting hemp crops are not yet available.

Practices regarding the organic fertilization of hemp have changed. Before World War II, farmyard manure was the most widely used hemp fertilizer. As a result of the present practices of the livestock industry, liquid manure is now used in place of farmyard manure. Hemp does well with liquid manure, but it is advisable to adjust the quantity used. The fertilizer should have a nitrogen equivalent of 100–150 kilograms N per hectare (90–135 pounds per acre), a phosphorus equivalent of 50 kilograms P_2O_5 per hectare (45 pounds per acre), and a potassium equivalent of 90 kilograms K_2O per hectare (80 pounds per acre).

If necessary, the liquid manure can be supplemented with a small amount of NPK fertilizer, according to the nutrient condition of the soil. Only a small portion the liquid manure is water-soluble and therefore immediately available to plants. Liquid manure contains high numbers of microorganisms that convert the unavailable nutrients to forms in which they can be absorbed by plants. In addition to its other advantages, liquid manure thus provides a continuous supply of essential nutrients to the plants. This is of the greatest benefit to hemp.

In August and September, the liquid manure can be brought out into the field in equal amounts, but care should be taken to avoid loss of nutrients by leaching. For this reason, the manure should be applied either in fall to an intermediate rotational crop, or in spring at the beginning of the vegetation period—directly to the hemp plants.

Regardless of when liquid manure is applied, prompt tillage is imperative. By application in the fall, the liquid manure rapidly begins to break down. Although the fall frost ends microbial life in the soil, which only resumes its activity in spring, nutrients supplemented with mineral-rich fertilizers are available to the plants at the time of germination. Recent studies from Hungary indicate that the beneficial effects of the usable nutrients from liquid manure last as long as two years.

5.2.1 Fertilization Methods and Timing

▓ *Nitrogen*

Nitrogen fertilizer enriched with phosphorus and potassium has the most positive effect on crop yield. However, excess nitrogen fertilizer is unfavorable for the crop, in particular for the fiber quality. In determining the amount of nitrogen fertilization necessary, the beneficial effects of the previous crop in rotation have to be considered. If the previous crop has produced a considerable amount of plant and root residue in the field, the supplemental amount of nitrogen fertilizer should be reduced. Typical amounts for nitrogen fertilization of hemp crops are 8–10 kilograms (18–22 pounds) of nitrogen per ton of dry matter. The current practice in Hungary is to apply half of the nitrogen fertilizer in the fall, before sowing of the hemp crop and after harvesting the previous crop in rotation. The second half is then applied in spring.

▓ *Phosphorus*

One of the primary characteristics of phosphorus is that it rapidly dissolves and is absorbed into the soil. Due to these properties and since it is readily available in granular form, phosphorus should be introduced directly into those soil layers that have the highest amount of root mass. Therefore, it should be applied and plowed after harvesting of the previous crop.

■ *Potassium*

The timing and the method for distributing potassium fertilizer are similar to those for phosphorus fertilizer. Since migration of potassium in the ground is minimal, leaching is not of concern. Potassium acidic with sulfuric acid is recommended for hemp cultivation, because chloride ions are detrimental to fiber development. As with phosphorus, potassium fertilizer should be applied after harvesting the previous rotational crop in the fall.

5.2.2 Fertilizer Quantity

The guidelines for fertilization are based on the determination of the active ingredient in fertilizers, on the results from scientific research, on practical experience, and on soil-testing results.

The nutrient content and other agricultural characteristics of a soil can be assessed through soil analyses. The organic matter content and the aluminum-soluble phosphorus and potassium content (i.e., NPK supply) can be classified as follows:

- ■ Poor
- ■ Low
- ■ Average
- ■ Good
- ■ Excellent

Within these categories, the organic matter content is placed into subcategories based on the aluminum soluble potassium content, the soluble phosphorus, and the $CaCO_3$ content (1% $CaCO_3$ absorbed, or no $CaCO_3$) in the topsoil. The nutrient content must be determined for every chemical element separately. In addition, the specific requirement of hemp for the specific nutrient must also be considered. Table 10 serves as a guideline.

Based upon these findings, the amount of fertilizer required for the anticipated yield can be theoretically determined, but the amount effectively used should always be adjusted to the actual conditions.

The following example illustrates the calculation method for the cultivation of hemp. A field with black soil from soil quality class 1 and a yield category (5) with a stalk yield between 8.0 and 9.0 tons per

Table 10: Specific fertilizer requirements of hemp,
depending on soil quality

Soil Quality	Fertilizer Requirements kg/metric ton stalk yield (lb/short ton)				
	Nutrient content of soil				
	Poor	Low	Average	Good	Excellent
NITROGEN					
Class I	21.5 (43.0)	18.0 (36.0)	15.0 (30.0)	14.0 (28.0)	8.0 (16.0)
Class II and III	23.0 (46.0)	20.0 (40.0)	18.0 (36.0)	15.5 (31.0)	10.0 (20.0)
PHOSPHORUS					
Class I	12.0 (24.0)	10.0 (20.0)	9.0 (18.0)	8.0 (16.0)	4.0 (8.0)
Class II and III	17.0 (34.0)	14.0 (28.0)	12.0 (24.0)	10.0 (20.0)	7.0 (14.0)
POTASSIUM					
Class I	24.0 (48.0)	18.0 (36.0)	10.0 (20.0)	12.0 (24.0)	7.0 (14.0)
Class II and III	27.0 (54.0)	24.0 (18.0)	22.0 (44.0)	20.0 (40.0)	8.0 (16.0)

hectare (3.6–4.0 short tons per acre) are assumed (see Table 7). The anticipated stalk yield is 8.5 metric tons per hectare (3.8 short tons per acre). According to the soil analyses, the $CaCO_3$ content is 2.1%.

■ Calculation

Determination of nutrient supply category according to nutrient content and soil quality class (refer to Table 8 and 9):

Phosphorus = 70 ppm P_2O_5,   Class I = poor supply
Potassium = 280 ppm K_2O,   Class I = average supply

Specific demand for nutrients according to nutrient content of soil quality class (from Table 10):

Nitrogen = 18.0 kg N/metric tons hemp straw
(36.0 lb/short tons hemp straw)

Phosphorus = 10.0 kg P_2O_5/metric tons hemp straw
(20.0 lb/short ton hemp straw)

Potassium = 10.0 kg K_2O/metric tons hemp straw
(24.0 lb/short ton hemp straw)

Total nutrient demand per acreage for an anticipated stalk yield of 8.5 metric tons per hectare (3.8 short tons per acre):

8.5 t/ha × 18.0 kg N/t = 153 kg N/ha (337 lb/acre)
8.5 t/ha × 10.0 kg P_2O_5/t = 85 kg P_2O_5/ha (187 lb/acre)
8.5 t/ha × 10.0 kg K_2O/t = 85 kg K_2O/ha (187 lb/acre)

If the previous crop was a legume (a nitrogen-fixing plant), the quantity of nitrogen can be reduced.

Nitrogen = 112 kg N/ha (245 pounds per acre)

The appropriate amounts of nutrients need to be adjusted according to the nutrient content of the soil.

5.3 Crop Rotation and Self-Tolerance

At the beginning of the 1980s, hemp crops accounted for 3–10% of the acreage of farms cultivating hemp in Hungary. In most cases, fiber hemp is cultivated in crop rotation between two cereal crops. Annual or perennial legumes are seldom used as a preceding crop. Fiber hemp has no demands regarding the preceding crop; nonetheless, it requires good soil conditions, plenty of nutrients, and excellent water content in the soil. A high stalk yield of good quality can be expected if the soil has been conditioned expertly and in a timely fashion after harvesting of the cereals crops or legumes during the previous year, and if an adequate supply of nutrients is available. Fiber hemp is one of the most beneficial crops to plant before cultivating cereal crops. Farmers planting winter wheat after hemp frequently experience yield increases of 10–20%. Yet fiber hemp currently is not an important preceding crop to winter wheat, which is cultivated on vast acreages compared to hemp.

Under present conditions, the harvesting of fiber hemp occurs continuously. The field is available to other crops starting between mid-August and mid-September. Fiber hemp leaves the soil in excellent condition; as a result of its ground-shading cover, the soil retains moisture and weeds are eradicated. Thus, following a crop of fiber hemp the soil is easy to work. After the hemp stalks have been harvested and transported, fertilizer may immediately be applied to the soil, the soil may be tilled, and the seedbed may be prepared for the following winter cereal crop.

Hemp is exceptionally self-tolerant; it can be cultivated two or three times in a row without resulting in significant yield loss. In such cases, however, the second and third hemp crops require more fertilizer and it is advisable to keep an eye out for pests (particularly the hemp flea). A hemp monoculture is not beneficial because there are so many suitable preceding crops available—primarily cereals—and because the high value of hemp as a preceding crop is not attained with a monoculture.

5.4 Tillage

Fiber hemp places great demands on the soil and its water-holding capacity, aeration, and heat retention. It is wise to choose tillage methods that retain precipitation, incorporate the nutrients into the soil, sustain porosity, and keep a smooth surface. Such soil conditioning enhances biological processes, thereby contributing to the nutrient cycle.

In Hungary, tillage begins with plowing after fall harvesting. Furrowing can be done either with a disk or with another kind of plow cultivator. In both cases, the topsoil should be packed and smoothed over with a roller. The porosity and heat-retention properties of loosely packed, smoothed soil are favorable. In addition, biological processes are then dynamic and evaporation can be kept to a minimum. If weeds begin to grow after the field has been plowed, retilling of the soil is recommended. Hard, heavy soil hinders the adequate preparation of the topsoil and becomes dry at sowing time, so a deep plow in fall is recommended. Otherwise the soil will become friable with the onset of frost. Either at the same time as plowing or immediately thereafter, packing the soil with a ring roller is beneficial.

September is the optimal deep-plowing time in eastern Europe. In Germany, this period can be moved to the end of October or beginning of November because of fall rains. Under no circumstances is allowing a field to lie fallow without vegetation recommended, for this inevitably results in erosion and nutrient loss. Rotation cropping, already customary practice in Germany, is recommended where the climate permits.

Table 11 shows the effects of the timing of fall plowing on stalk yield, as determined in Hungary. The results clearly demonstrate how careful, thorough soil preparation significantly promotes the overall yield of a hemp crop.

Table 11: The effects of the timing
of plowing on hemp-stalk yield

Month of fall plowing	Stalk yield	
	metric tons per hectare (short tons per acre)	%
Soviet data		
September	4.9 (2.0)	100
October	4.1 (1.8)	84
November	3.8 (1.7)	71
Romanian data		
September	5.8 (2.6)	100
October	4.6 (2.1)	80
November	4.0 (1.8)	69

On sandy soil, fall plowing is not carried out in Hungary. Fiber hemp is planted relatively early in the season—at about the same time as sugar beets. Before sowing, the timing and method of soil preparation should be chosen in such a way that the soil condition promotes rapid germination and strong seedling growth.

Fertilizer should be uniformly applied early in the year, before sowing, so it can be incorporated into the soil during preparation of the seedbed and sowing. When tilling the seedbed, tools should be used that loosen the soil and destroy the early weeds without mixing it and turning it clumpy. The fine, friable soil condition should be preserved. The soil should not be tilled any deeper than one intends to sow. With dry climates and loose soil conditions, it is advisable to smooth the soil and pack it with a roller. The seedbed for sowing hemp must have equally fine, friable soil from the surface down to sowing depth.

5.5 Sowing

The sowing period is dictated primarily by spring weather conditions. Although hemp seeds begin germinating at temperatures ranging from 1–2 C (34–36 F), hemp is not an early-sowed plant. In order to benefit from hemp's rapid growth, which facilitates the plant's ability to outcompete weeds, it is recommended to wait for sowing until the soil temperature has reached 10–12 C (50–54 F). In southern and central Germany, such temperatures are normally reached in the last ten days of April, and in northern Germany during the first ten days of May. Experience shows that even the southern varieties then accumulate enough heat, amounting to 1,800–2,000 GDD_C (3,300–3,600 GDD_F) by the last ten days in August or the first week in September, which is necessary if the plants are to reach technical maturity. Therefore, even in northern Germany, the latest time to sow hemp is the twentieth of May. In Germany, the optimal sowing period lasts three to four weeks (from the end of April to the end of May).

It is advisable to sow the early-maturing varieties near the end and the late-maturing southern varieties at the beginning of this sowing period. For future cultivation in Germany, it will be necessary to conduct sowing tests in weekly intervals during the sowing period (from the middle of April to the end of May) in various soil types and climatic zones, because the optimal sowing period is dependent on the specific cultivation area. Because spring weather conditions can be quite different from year to year, a minimum of ten years of testing will be necessary before practical conclusions can be drawn.

As a rule of thumb, any machinery is suitable for sowing, as long as the spacing of the seeds in row widths of 12–20 centimeters (4–8 inches, the same as for wheat) is adjustable. Any seed drill that is easily adjustable and performs well is suitable for sowing hemp seeds.

The seed depth is generally 3–4 centimeters (1.2–1.6 inches), and only slightly deeper in sandy soil. If, in mineral soil, seeds are sown deeper than 3 or 4 centimeters (1.2–1.6 inches), they will germinate unevenly. The sprouts will have a difficult time reaching the topsoil, and some of the seedlings will die. If the seeds are sown too shallowly, the seedbed will become too dry and the plants will germinate irregularly. Table 12 illustrates the effect of sowing depth and thousand grain weight (TGW) on seedling yield (expressed as a percentage). The deeper the seeds are sown, the lower the plant yield. A low thousand grain weight is only important at a depth of 5–7 centimeters (2–2.8 inches).

Table 12: Hemp yield dependent on
sowing depth and amount of seed

TGW g (ounces)	Seed depth		
	3 cm (1.2 in)	5 cm (2 in)	7 cm (2.8 in)
20 (0.7)	79%	50%	47%
15 (0.5)	75%	41%	40%
12 (0.4)	73%	28%	20%

The correct sowing technique significantly affects overall yield. Fiber content and quality are dependent on genetic characteristics, particularly in southern and transitional varieties. In addition, fiber content and quality are influenced to the extent that the length and thickness of the fiber stalk are correlated. The longer and thinner the stalk, the higher the fiber content and the better the fiber quality. The desired development of the stem can be attained through adequate preparation of the field as well as proper plant spacing (see Table 13).

In Hungary many experiments on plant growth have been conducted. A number of large farms conducted experiments on their cultivated fields in 1977 and 1978. Later, experiments were carried out in Kompolt on experimental plots (1995).

As Table 13 shows, the results of tests conducted in large fields and small plots completely concur. Significant differences for stalk yields between 60–100 kilograms of seed per hectare (55–90 pounds per acre) or 300–500 plants per square meter (28–46 plants per square foot) were not found in either of the tests. In the experiments conducted on large fields, the difference between the two extreme values was only 600 kilograms of stalk per hectare (530 pounds per acre), which is not convincing, primarily because this yield increase is offset by a higher seed requirement of 20 kilograms (44 pounds).

In any case, the tests clearly demonstrated the self-regulating capacity of hemp. This means that the yield cannot be further increased once a certain seed quantity and plant density per acre or hectare has been reached. The hemp that develops from the excess plants will have a stalk height of only about 50% of average. In the traditional process of fiber separation, tow is produced from shorter plants rather than long-fiber ones. Tow cannot be used in the spinning industry.

According to results from the latest experiments conducted on small plots, the optimal economical seed quantity is 70–80 kilograms per hectare (60–70 pounds per acre). Under no circumstances should the grower exceed a seed quantity of 80 kilograms per hectare (70 pounds per acre). The results clearly demonstrate that at quantities greater than 80 kilograms per hectare (70 pounds per acre), the amount of hemp of insufficient height, which is worthless to the hemp grower, rapidly increases. In other words, a seed quantity exceeding 80 kilograms per hectare (70 pounds per acre) is simply a waste. Contrary to present

Table 13: The effect of seed quantity on stem yield
in large fields and small plots

Seed count 10^6/ha (10^6/acre)	Sowing strength kg/ha (lb/acre)	Plant count (harvest) 10^6/ha (10^6/acre)	Plants of insufficient height % of total crop	Stalk yield metric tons/ha (short tons/ acre)
LARGE FIELDS TÁRKÁNY-SZÜCS 1981				
3.5 (1.4)	70 (62)	3.1 (1.3)		8.36 (3.73)
3.8 (1.5)	76 (68)	3.6 (1.5)		8.40 (3.75)
4.0 (1.6)	80 (71)	3.6 (1.5)		8.52 (3.80)
4.5 (1.8)	90 (80)	4.1 (1.7)		8.96 (3.98)
5.0 (2.0)	100 (89)	4.7 (1.9)		8.91 (3.97)
SMALL PLOTS (KOMPOLT 1995)				
3.0 (1.2)	60 (54)	1.54 (0.62)	14	9.57 (4.27)
3.5 (1.4)	70 (62)	1.98 (0.80)	11	9.23 (4.12)
4.0 (1.6)	80 (71)	2.14 (0.87)	15	9.68 (4.32)
4.5 (1.8)	90 (80)	2.42 (0.98)	19	9.45 (4.21)
5.0 (2.0)	100 (89)	2.57 (1.04)	23	9.23 (4.12)
10 m² (108 square feet), 5 repetitions			LSD$_{5\%}$*	0.96

*LSD$_{5\%}$ = least significant difference at p=0.05

practices, it would even be advisable to reduce the quantity to 70 kilograms per hectare (60 pounds per acre) or about 3.5 million plants per hectare (1.4 million plants per acre).

Hemp that is being grown for paper production requires less seed than hemp grown for the textile industry. Row intervals of 20 centimeters (8 inches) and 45–60 kilograms of seeds per hectare (40–55 pounds per acre) result in 2.2–3 million sprouts per hectare (0.9–1.2 million plants per square meter), which is sufficient. Fewer seeds would also be adequate, but the stalks would become thicker as a result of larger row intervals. The thicker stalks would impede or completely prevent the processing of the stalks with a reaper and bale press.

The seeds should have a water content of 12%, a purity ranging from 98.5–99.5%, and a germination potential of 85–90%.

The potentially hemp-damaging weeds, mostly dodder (*Cuscuta europaea*) and broomrape (*Orobanche ramosa*), are rarely found in hemp seeds. Seeds should definitely possess the qualities described above at the time of sowing, because the germination potential of hemp seeds rapidly declines. It is important to determine the germination potential of carry-over seeds, because the potential may fall below 80% if the seeds were improperly stored. Seeds properly stored in refrigeration units will still be suitable for sowing after three years. Seeds stored at a temperature of –10 c (14 F) can even germinate after a period of five to six years.

The topsoil usually cracks open after rapid vaporization or extensive rainwater seepage. If this happens immediately after sowing, the cracking process hampers germination, or leads to irregular plant growth. When cracks appear this early, it is absolutely necessary to clear them away with a roller or hoe.

5.6 Plant Protection

The most significant damages inflicted upon fiber hemp crops can be categorized into the following groups:

- weeds and parasitic plants
- insects and other animal pests
- fungal diseases
- herbicide damages
- weather damages

5.6.1 Weeds and Parasitic Plants

Due to its outstanding ability to suppress weeds during its vegetative period, hemp requires no herbicides during that time. Weeds are eradicated early in the year, during soil preparation for sowing. Hemp plants overshadow the soil fairly soon after their initial growth phase, which means that chemical herbicides are unnecessary. Weed control by mechanical means cannot be done after the plants have sprouted without causing damage to the plants. Any temporary patches of weeds soon disappear once the hemp plants begin to overshadow them.

Among all weeds and parasitic plants, only one is of importance to hemp cultivation:

■ **Broomrape (Orobanche ramosa)**

Broomrape is a highly parasitic plant that damages fiber hemp. However, it appears only with Chinese varieties in seed-bearing stock, which are not cultivated in Germany or western Europe. *Orobanche ramosa* is about 15 centimeters (6 inches) tall or larger, with fine-haired ochre-yellow to brownish shoots. Since it is a parasite, its roots penetrate hemp roots and extract nutrients. Modern seed-cleaning methods have significantly reduced the risk of damage from broomrape, yet the cultivation of resistant hemp varieties is recommended nevertheless.

5.6.2 Insects and Other Animal Pests

Recently, many people have erroneously claimed that hemp does not suffer damage from animal pests. This, however, is not in accord with the facts, although none of the recognized pests usually causes economic losses. In central and southeastern Europe, hemp has several recognized insect pests. It is not yet known how these and other pests will adapt to the increasing hemp production in western Europe.

■ **Hemp Flea (Psylliodes attenuata KOCH)**

The hemp flea is the most dangerous insect pest to industrial hemp. The damages are noticed only when the temperature of the top layer of soil and the air temperature have increased to 10–15 c (50–60 f) and the weather is dry. The hibernating fleas then crawl from the soil and feed on the cotyledons as well as on the hypocotyl (the seedling stem

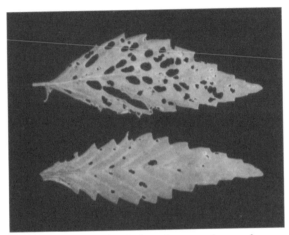

Figure 19: Hemp flea infestation; *top:* heavy damage
on a susceptible variety; *bottom:* minor damage on
a resistant variety (southern × Chinese hybrid)

below the cotyledons). If the plants are already developed, the fleas will even chew small holes in the first leaves (see Figure 19).

If the hemp fleas have reproduced extensively, they can consume the complete hypocotyl from below the cotyledons down to the topsoil. More than 50% damage to the surface of the leaves can kill the plants. The plants can survive less damage, but they will develop at a slower rate and the overall crop yield will be reduced. Having to fight the hemp flea is somewhat unpredictable, because these pests are not a threat every year.

If noticeable damage to the plant exceeds 20%, pesticides can be used. For this purpose, a 50% solution of methyl parathion in 0.5–0.7 liters per hectare (7–10 fluid ounces per acre) is recommended. Since the hemp flea hardly ever occurs in numbers large enough to threaten entire crops, it is rarely necessary to adopt such extreme measures.

■ *Hemp Borer (Grapholita delineana)*

Hemp borers, also known as hemp moths, first proved to be a problem in the late 1960s. This pest has two or perhaps three generations per year. The larvae from the first generation damage fiber hemp. During winter, this generation hibernates in the stalk and stubble. Early the following year (after mid-May), when the temperature reaches 15 C (60 F), the larvae spin cocoons around themselves. After one or two weeks, the moths are fully developed and emerge from the cocoon. Each moth is 6–7 millimeters (0.2–0.3 inches) long and greyish-brown, and has three to four white stripes on its back.

The hemp moth is a skilled flyer, and it searches for fiber hemp fields by taking advantage of its excellent sense of smell. The female lays one hundred to two hundred eggs, scattering them beneath the growth apex of the stalk, i.e., on the youngest part of the plant. The tiny hatching caterpillars feed on the back side of the leaves during the first larval stage. In the second larval stage, after pupation in the ground, they feed in the cavity of the stem medulla. The caterpillars continue to develop here until they reach the fifth larval stage, or until the wood tissue has developed to the point that the larvae feed on it by chewing the wood with their teeth. They proceed to crawl through the hole and begin feeding on the younger, more tender sections of the stalk. In the fifth larval stage, the larvae spin cocoons around themselves and develop into moths. This generally occurs at the end of June or beginning of July. The damage that borers do is visible as trails on the stalk, and also often as thicker sections of the stalk, which are referred to as galls. Such hemp stalks have a low fiber yield, and the quality does not meet industry standards.

The fight against the hemp borers begins shortly after mid-May with periodic inspections of the plants. At dusk, it is advisable to use nets on the perimeter of the fiber-hemp field facing the hemp field from the previous year. An average catch of one half to one moth per ten swings indicates that it may be time to use an insecticide. If so, it is best to start with an orchard sprayer from the edge of the new field (the sprayer cannot be taken into the hemp field!). The orchard sprayer has a range of 8–10 meters (26–33 feet), which is sufficient to spray the perimeter of the field (using, for example, a 50% solution of methyl parathion in 1.4–1.6 liters per hectare, or 19–22 fluid ounces per acre).

The second generation of hemp moths appears on the fiber hemp at the end of July. Since harvesting has already begun, the larvae of this generation cannot fully develop on the plants. The larvae move from the harvested, drying hemp and crawl into the moist soil. The larvae in the fourth and fifth stage spin themselves into cocoons in which they hibernate. In Hungary, the farmers burn the stubble and the stalk remnants after the harvest, as a preventive measure. The field is plowed with a deep furrow, and the wild-growing hemp along fields and roads is destroyed. In Germany, the burning of the stubble and the stalk remnants is prohibited because of the danger of uncontrolled fires, heavy air pollution, and damaging effects to wildlife.

■ *European Corn Borer (Ostrinia nubilalis)*

Damage to fiber hemp crops by the European corn borer occurs primarily in eastern and southeastern Europe. The last such damages worth noting were registered in the late 1950s. Damage occurs in the form of bore holes created by the caterpillars as they bore through the stalk, but no gall is formed. Since the European corn borers are larger insects and consume more than the hemp borers, they devour larger pieces of wood, and the hemp stalk then breaks in the wind. Preventive measures have been adopted in an attempt to keep European corn borers from reproducing. A system for fighting these pests with the use of chemicals has never been generated, but if the European corn borer multiplies, the methods used against the hemp borer (*G. delineana*) appear to be promising.

■ *Other Pests*

Fiber hemp can also be damaged by hemp greenflies (*Phorodon cannabis*), but this pest has never occurred in large numbers.

The northern rootknot nematode (*Meloidogyne hapla*) has recently been reported in the Netherlands. These pests can cause significant damage to crops in countries with an oceanic climate, such as Belgium, Luxembourg, the Netherlands, England, and western sections of Germany. The nematodes can presently be controlled only by using resistant plant varieties.

5.6.3 Fungal Diseases

■ *Pythium Disease (Pythium debaryanum)*

This fungus, commonly referred to as "the young plant disease," attacks both the seeds and the sprouting plants. It is a nonspecific parasite and attacks many different varieties of sprouting plants. The pathogen breeds in wet, poorly aerated soil. The disease can be combatted with appropriate agrotechnical actions. The fungus mainly appears during the germination stage or shortly thereafter, and causes the plants to topple over.

■ Hemp Canker (Sclerotinia sclerotiorum)

This disease occurs in wet soil or during years of particularly high amounts of precipitation. The fungus attacks the plant tissue at the base of the root and eventually kills the plant. Damage to the plant occurs in the form of a yellowish-brown discoloration on the lower portion of the stalk. Later, the leaves turn yellow and fall off, and the plant dries out. The best method of control is to avoid fields with high water tables and to improve the water balance of the soil by means of technical cultivation methods.

■ Grey Mold (Botrytis cinerea)

The affected plants in the crop develop blotches, and the stalks then break and topple over. Grey mold can also affect the crop during field retting if the stalks get wet repeatedly due to recurrent rainfall. Presently, no method has been developed to fight grey mold, which causes crop damage primarily in western Europe. It is possible to fight grey mold by means of resistance breeding, crop rotation, and soil rejuvenation. This disease is practically nonexistent in the dry climates of southeastern Europe.

■ Hemp Rust (Melampsora cannabina)

The presence of this disease can be detected when, on both sides of the leaf, orange-colored blotches appear, from which yellow spores fall. The pathogen attacks the plant fiber. The control method is to spray with a solution containing thiocarbamate (Zineb-Maneb). Specific conditions must be met when using Maneb, because this chemical has a high aqueous toxicity and fish toxicity. A minimum distance of 20 meters (65 feet) has to be kept from lakes, rivers, or streams. Furthermore, Maneb cannot be used in any areas where there is a significant danger of runoff or erosion. Hemp rust does not result in infestation epidemics.

As far as hemp is concerned, the specified diseases are all rare as epidemics. Specialized literature describes a number of other pathogens in addition to those mentioned here, but these are rarely, if ever, a threat to hemp cultivation.

5.6.4 Herbicide Damage

■ *Triazine Derivatives*

Damaging effects caused by residue from triazine derivatives (i.e., Atrazine) have occasionally been detected on smaller plots. In every case, the damage first manifested itself in the crop during the first growth stage (plant height of 5–15 centimeters, or 2–6 inches), and first on the more developed plants, in the form of drying leaves and deformation of the stalk. Such a plant ultimately dies. This process first occurred in patches, then in smaller sections of the crop. The soil in these areas was found to contain triazine derivatives even after the stipulated waiting period had elapsed, which shows that hemp is extremely sensitive to residue from triazine derivatives. This residue also regularly appears when the previous rotational crop has been corn that was repeatedly treated with triazine.

■ *Leaf Herbicides*

In several farms in Hungary during the past few years, deformed leaves, petioles, and stalk portions have occurred with greater frequency in fiber-hemp crops in which plants had already reached a height of 40–60 centimeters (16–24 inches). The damage could be shown to be a direct result of the use of leaf herbicides applied with airplanes onto neighboring cereal crop fields. In all of these cases, the prescribed techniques for distributing the herbicides on cereal crop fields were obeyed. Nevertheless, air movement caused by the planes transported very small droplets of herbicide spray to the neighboring fiber-hemp fields, and resulted in all of the fields being affected.

In smaller quantities, the derivatives of 2,4 dichlorophenoxyacetic acid (2,4-D) do not completely destroy hemp plants. However, the leaves, petioles, and stalks become deformed, growth is halted, and intensive growth does not resume until two to three weeks after. The stem segments (internodes) branch out like a fork and the nodes become elongated (see Figures 20 and 21). Such stalks no longer have any industrial value, and the deformed stalks are classified as damaged.

Similar experiences with hemp's herbicide sensitivity have also been referenced in Germany. For example, beet breeders who have grown hemp as a buffer crop in their breeding fields have noted its sensitivity to herbicides.

The damage caused by herbicides can be avoided with proper handling and application. However, any herbicide damage done is irreversible; the plants do not "grow this damage out."

Please note: The licensing status of some of the agricultural chemicals mentioned throughout this book will vary from one country to another. Readers should consult their local agencies.

Figure 20: Plant damaged by 2,4-D herbicides

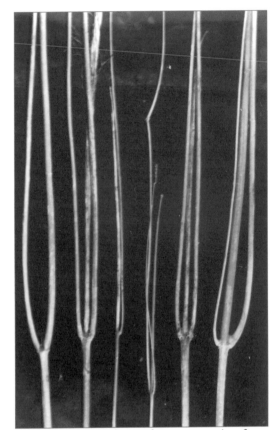

Figure 21: Hemp plants after regeneration from
2,4-D damages: fork-shaped branching of stem

5.6.5 Weather Damage

■ *Hail*

Falling hail can cause severe damage to a hemp field. Even small hailstones can break the stalks of young, developing plants of 40–60 centimeters (16–24 inches) in height (in Germany, from May to June). The plants either die or develop secondary shoots that are not capable of surviving. The stand is thinned out by the number of dead plants, the spacing between the remaining plants increases, the fiber hemp stalks become thicker, and their quality worsens. If branches develop on a stalk, that plant is no longer suitable as fiber hemp. If the crop becomes so thin that the average diameter of the remaining developing plants or stalks exceeds 15 millimeters (0.6 inches) or if the female plants branch out, the crop is considered a total loss. At this point the crop should be plowed under.

Damage to the stalk may also result if the hemp plants are struck by hail after the lower stalk section has lignified. It does not matter whether the hailstones are small or large, round or angular, occur with or without rain, or are accompanied by heavy or light winds. In each case, the delicate growth apex can be damaged and can break. When the sprouts are broken but the broken part still remains on the stalk, the remaining portion can survive. The attached portion of the sprout is bent downward, but will eventually grow upright again. Secondary branches develop on the broken stalks, which lead to inferior quality.

The injuries caused by hailstones heal very slowly. The wounds are frequently infected by fungal pathogens (secondary damage). The extent of the damage is determined by the suitability of the fiber for further uses, which is in turn determined by the changes in the stalks of those plants still capable of producing fiber. If the wound impacts not only the epidermis but the bast, or the wood tissue, the outside dries out after a few days, hardens, and becomes cork tissue (as long as no fungal infection occurs). This cork tissue is the external evidence of damage. The associated changes to the plant tissue however, are more significant than the external damage. At the point of impact, the plant cells and tissues are damaged and the plant immediately balances the damage to its supporting stroma and strengthening tissues by fortifying the tissue fibers on the opposite side of the stalk. The quality of the fiber tissue is reduced in the affected area. During industrial processing, the fibers splinter at the damaged areas. For this reason, all externally damaged areas on the hemp stalk should be meticulously evaluated, regardless of whether the stalks were damaged by hail or through other circumstances.

Only fiber-hemp crops are damaged by hail. Hail damage does not result in losses in hemp that is cultivated for its cellulose (paper). Minor damage occurs in widely spaced hempseed crops, because the injured plants branch out and produce more seeds. However, more densely spaced seed crops can suffer damage from hailstorms.

■ *Wind and Storm*

As long as mowing and handling of the stalks was done manually, wind or storm damage was not classified as real damage; the broken hemp stalks were simply thrown away. But mechanized harvesting makes such manual selection impossible.

The effects of wind can cause fiber hemp stalks to come together, scrape against one another, and even break. If the stalks are lying in the field in such disarray that the gathering device on the mower is unable to put them in order, mechanized harvesting should be abandoned. In such cases, if harvesting by hand is not possible, the crop should be destroyed. The industrial assessment of broken and damaged stalks from wind and storms is identical to damage assessment following a hailstorm.

■ *Drought*

Drought can reduce the yield of a fiber-hemp crop just as it can lessen the yield of any other cultivated plant. The hemp stalk remains smaller, and if the stalk length is near a quality-class limit, it is likely that the fiber quality will be classified in a lower group.

■ *Water*

Fiber hemp should not be cultivated near inland waterways or in areas prone to high water tables, because hemp will not survive if it is exposed to excessive amounts of water (moisture buildup) for more than twenty-four hours. The same holds true for long, hard rainfall, if the water remains in the field for longer than twelve hours. In such a case, the development of the plant may be halted, so that it turns yellow and may eventually die.

Chapter 6

■ ■ ■ ■ ■ ■ ■

Harvesting

Harvesting is presently the most problematic and least understood part of hemp cultivation. Currently, two harvesting techniques have been developed and are being employed. The first of these techniques is the French harvesting method, which is applied to hemp cultivated for the paper industry. This method is well-developed but cannot be uniformly implemented and standardized because of its single- and dual-purpose forms (harvesting seeds and fibers either individually or in the same field), and because it is the farmer who ultimately decides which form to adopt.

The second hemp-harvesting technique is the one used for the textile industry. This method has been adopted by countries formerly belonging to the Soviet Union as well as other eastern European countries. Depending on the country, the method may incorporate few or many variations. One common factor, regardless of variations, is that this harvesting technique is based upon use of a Soviet-built combined mower/binder (SchSK 2.1).

In Hungary, the RKB-1 bale-press machine was developed to further simplify the harvesting process. The RKB-1 presses the hemp plants into bales weighing 400 kilograms (900 pounds), thereby making loading and transport significantly easier. We have been unable to ascertain whether the Russian firm still produces or, for that matter, exports the SchSK 2.1 mower/binder. Moreover, this machine was suitable for harvesting large crops only after modifications were made to the binding machine and the mowing mechanism. The need for such modifications made it necessary for the Hungarian government to establish repair facilities where the work could be done.

If the hemp-textile industry enters a substantial upswing in western Europe, the need for a new, modern harvesting technology from some country in the European Union (optimally, from the German mechanical-engineering industry) will surely be met. In the meantime, the hemp industry is forced to make do with the current Russian (formerly Soviet) machine; with the Hungarian bale press, based on technology adopted from the textile industry; and of course with the advanced French harvesting method for the industrial paper industry. The choice of which harvesting technique to adopt is dependent upon the utilization goals:

- ∎ technology for long fibers;
- ∎ technology to obtain the whole fiber as a universally usable short fiber;
- ∎ technology to obtain seed and short fiber simultaneously (dual-purpose).

6.1 Eastern European Technology

In eastern Europe, where industrial paper technology is nonexistent, fiber hemp is harvested for the textile industry. The harvesting is mostly done with Russian machinery and methods. Hemp was—and perhaps still is—harvested for dual purposes in central Russia and a small area in northern Ukraine, but this shall not be considered here.

6.2 Hungarian Technology (Fiber Hemp)

6.2.1 Determination of the Harvesting Date

The point of harvest is crucial for the quantity and quality of the yield as well as the fineness and content of the fiber.

The deciding factor in determining when to harvest is that the fiber is available in the largest quantity and is of the best quality. With male plants, this is during flowering; with female plants, the ideal point is at the first appearance of flowers. Therefore, literature describes this particular stage of development as the point of technical maturity—in other words, the best time to harvest. Thus, harvesting of fiber hemp in agricultural production should begin when one-third of the anthers in the male flower are open and shedding pollen. Harvesting should be completed before the male plants have stopped flowering.

In Hungary, the optimal time to harvest is generally from the last days in July through mid-August, depending upon the weather (dry and sunny versus cool and rainy). In central and southeastern Europe, the early-maturing monoecious and dioecious varieties grown between 50 and 55 degrees north latitude should be harvested between mid-August and the first ten days of September. Hemp crops harvested any later will not have time to adequately dry in the field. Within this twenty-day period, the early-maturing varieties should be harvested in mid-August and the late-maturing varieties in early September. In the interest of lengthening the harvest period, it is advisable to grow at least two varieties with different points of maturation.

The harvesting of hemp currently consists of six phases from a process developed in Hungary during the 1970s:

1. chemical defoliation
2. reaping with a sheaf binder
3. stacking
4. baling
5. loading
6. transport

6.2.2 Chemical Defoliation

The leaf mass of air-dried hemp stalks (dried to a moisture content of 10–14%) accounts for 9–14% of the total plant weight. The leaves are an unnecessary burden, both during transport and in the retting basin. They take up a significant amount of space, result in additional costs, reduce the retting capacity in the retting basin, and impart an undesirable discoloration to the stalks. For these reasons, it is necessary to eliminate this unnecessary burden.

In Hungary, it was decided more than twenty years ago that defoliation by hand was so difficult and unproductive as to be completely out of the question. Yet attempts to mechanize the harvesting process have thus far been unsuccessful. Therefore, Hungarian hemp farmers have turned to chemical defoliation as a solution.

Diquat and paraquat defoliants were not considered, for they diminish the tensile strength of the hemp fibers and are also a health hazard. Purivel, a defoliant that is gentler on the fibers and contains the active substance Metoxuron (developed by Sandoz), is the substance that has been put to use. Larger hemp fields of 40–100 hectares (100–250 acres) have been considered particularly well-suited for chemical defoliation because they could be quickly and inexpensively treated with airplanes. Depending upon leaf coverage and spraying time, 4–6 kilograms (9–13 pounds) of the active ingredient in 60–100 liters of water per hectare (6.5–11 gallons per acre) have been used.

Purivel has a systemic effect; once absorbed into the epidermis of the leaves, it hinders the synthesis of chlorophyll, primarily in the new leaves. The leaves begin to dry out five to eight days after spraying. The edges become brown, and after ten to twelve days the leaves fall off.

Products equally effective and more environmentally friendly have recently been introduced. These newer defoliants are produced from natural raw materials derived from plants. Basta, manufactured by Hoechst Corporation, is an example of such a product. Basta contains the active substance ammonium glufosinate. The recommended quantity, which is dependent upon the total amount of leaves, is 2–3 liters per hectare (30–40 fluid ounces per acre). It should be applied when the male flowers are 10-30% open. A later application at the end of flowering is not effective. The active ingredient in Basta takes effect after four or five days. It does not translocate, nor does it block the synthesis of proteins. However, Basta definitely poses a health hazard. It is toxic to fish, and may only be used on fields from which the flowing time is at least fifty days to the next water treatment facility.

Another defoliant is Round-up, which contains 20% glyphosate. The main disadvantage with glyphosate is that it must be dispensed in significantly larger quantities than Basta. Round-up cannot be utilized closer than 10 meters (33 feet) away from any lake or stream because it is fish toxic. It is applied with 7–10 liters per hectare (95–140 fluid ounces per acre); the effects start within ten to fourteen days, depending on rainfall and temperature.

None of these chemicals can be sprayed by airplane on fields smaller than 25–30 hectares (62–74 acres), simply because this method is too expensive. For smaller fields, it is only worth spraying by airplane in cooperation with other farmers who farm adjoining fields. This approach is also advantageous to those farming larger acreages. Most airplanes have a 500- or 1,000-liter (130- or 260-gallon) tank and can operate only about four hours per day, early in the morning or in the late afternoon. Another disadvantage of using pesticides with an airplane is the difficulty of accurately dispensing and distributing the chemicals. Moreover, there is a considerable danger that the solution may drift due to minor weather disturbances or improper handling.

Theoretically, it is also possible to use an orchard spray gun on the tractor. This method can cover a distance of twenty meters (65 feet). If the spray gun method is adopted, lanes 2.5 meters (8 feet) wide will be required at 40-meter (130-foot) intervals, which will take up a total of about one-tenth of the acreage. In this lane area, potatoes, beans, or other plants that grow well in the shade can be grown, so as not to lose this land for production. Once the hemp plants reach a height of 1.5–2.0 meters (5–6.5 feet), these plants will be exposed to direct sunlight only at midday. The path for the tractor should be cleared no later than August 1. Due to a lack of practical experience, it cannot be determined whether this method is worthwhile, but testing is recommended.

Chemical defoliation of fiber-hemp plants takes place when the male plants have reached 10–15% flowering, roughly five to eight days prior to technical maturation. Any earlier application will result in yield reduction, and a later application may result in ineffective defoliation.

6.2.3 Alternatives to Chemical Defoliation

For traditional eastern European long-fiber crops, chemical defoliation is a method used to optimize quality and value. But since the collapse of the Soviet Union (the primary hemp market) and the simultaneous increase in demand for western European hemp for the production of "ecoproducts," defoliation has been dramatically reduced. The ecologically oriented textile manufacturers in western Europe will accept no goods that have been subjected to chemical defoliation, just as they refused to accept the previously common practice of PCP treatment of long fibers for protection against fungal diseases. (The latter practice has now been abandoned altogether.)

Modern, precise analytical techniques are used to provide evidence that no chemicals have been used during cultivation. Additionally, an increasing number of businesses are interested in certification of the hemp crop in accordance with the criteria for "certified organic farming." Certified goods consistently obtain a higher market price than do uncertified products.

What are the alternatives to chemical defoliation? For one, farming can return to methods used before the introduction of chemical defoliation. A principal example is manual/mechanical defoliation through brushing the leaves on the stubble, although this method is very time- and labor-intensive. Another option is to mechanically defoliate the plants after drying the hemp, but prior to water retting. In this case, new machines that preserve the parallel stalk orientation will be absolutely necessary. Finally, the hemp can be water-retted without being defoliated. This, of course, has economic disadvantages, including inefficient usage of the retting basin, increased transport volume (about 20% additional volume from the leaves), and undesirable fiber discoloration caused by chlorophyll from the leaves.

Newly developed and modified western methods for processing fiber hemp require no chemical defoliation, because with these methods water retting is no longer necessary—and water retting is the only process that benefits from defoliation of the stalks.

6.2.4 Mowing

In the former eastern European socialist countries, where hemp has been continuously cultivated, reaping was exclusively dependent on Soviet machinery. This was because only the Soviet industries were capable of manufacturing these machines in economically viable quantities. Thus the eastern European countries could be supplied with adequate machinery. These countries have made and still make exclusive use of the combined mower/binder SchSK 2.1, produced in the Beshetyk machinery plant. Figure 22 provides a picture of the SchSK 2.1.

A 50–80 horsepower tractor is necessary to pull the mower/binder. A driver and one other worker are normally required to operate the equipment. The new model of the SchSK 2.1 can be hydraulically adjusted from the cockpit of the tractor.

Eight to ten days after application of a defoliant, the hemp leaves turn brown and dry out, and 40–50% of the leaves fall off the plants in the wind. The male stalks turn yellow and the chlorophyll breaks down; the female plants die and their stalks turn yellowish-green. By this point, the stalks have already lost about 50% of their water content—a critical factor for the proper functioning of the binder device.

The SchSK 2.1 is relatively simple to operate. The separating arms, located above the cutting device, direct the stalks to the elastic bands. The cutter is 2.1 meters (6.9 feet) long, with a cutting width of 1.9–2.05 meters (6.2–6.7 feet). There are five separating arms and five pairs of rotating rubber belts opposite one another that hold the stalks while they are cut. The stalks are then transported over a glider plate to the binding device. In the sheaf-binding section of the binding device, the pressing equipment forms sheaves 20–25 centimeters (8–10 inches) in diameter and binds them with sisal or hemp twine. The binder device then ejects the sheaves onto the field. About 7–10 kilograms of binding twine is required to reap one hectare of fiber hemp (6–9 pounds of twine per acre).

Figure 22: The Russian SchSK 2.1 mower/binder

Under optimal conditions, the mower/binder can complete 5 hectares (12 acres) in a ten-hour period. An equal or better performance can be attained in central and southeastern Europe only with fiber hemp of a height of 200–240 centimeters (6.6–7.9 feet) which has been successfully defoliated.

Performance diminishes if the fiber hemp was not defoliated or if the defoliation was unsuccessful. If the hemp leaves are still green, the separating mechanism on the binder device may not be able to arrange the stalks as a result of high water content, and the machine will jam. For the reaping of plants with green leaves, the cutting width on the machine should be reduced. In addition, the binding mechanism should be arranged to form sheaves no larger than 15 centimeters (6 inches) in diameter.

6.2.5 Stacking

The sheaves should be placed in stacks immediately after the harvest is reaped (one stack equals about forty sheaves). Two to four mower/binders can simultaneously operate on a single field; six to eight machines may be required on larger fields. The mower/binders are efficient enough to keep other employees busy stacking the sheaves as they are ejected from the machines.

Fiber-hemp stalks lose about 50% of their water content between defoliation and reaping. By the time the bales are pressed, the plants should be completely dried out, with a water content of only 10–15%. The sheaves dry more slowly if left on the ground than they do when placed in stacks. A single stack can contain up to forty sheaves and still maintain acceptable air permeability. The stacks are placed in two rows separated by at least four meters (13 feet). Later, the stacks can be loaded two at a time.

One hectare usually contains between 1,600 and 2,000 sheaves (and one acre contains 650–800 sheaves), depending on the crop yield and the binder device's settings. This equates to forty to sixty stacks per hectare. Groups of workers are usually responsible for stacking the sheaves (two workers per stack). The workers can complete a total of 5 hectares (12 acres) in a day, which is equivalent to the time it takes for a mower/binder to harvest the entire field.

6.2.6 Press Baling

After the harvest has been reaped and the sheaves placed into stacks, the plants still need six to eight days to dry out. The water content should fall below 16%. Only dry stalks (with a water content below 16%) can be pressed into bales by the Hungarian RKB-1 bale-press machine (refer to Figure 23).

Figure 23: The Hungarian RKB-1 bale-press machine

The RKB-1 bale-press machine can be operated with the same tractor used with the combined mower/binder. Four workers are necessary to load the device. The pressing mechanism is attached to small wheels and has a ground clearance of only 100 centimeters (40 inches). Loading the hemp sheaves is thus not very difficult or physically exhausting. About eighty sheaves can be loaded into the pressing mechanism, which is the equivalent of two stacks or 400 kilograms (900 pounds). If the tractor operator and the workers loading the sheaves are experienced, the RKB-1 can bale 35–40 metric tons (39–44 short tons) of hemp in a single day.

In the former Soviet Union, the SchSK 2.1 is used not only for sheaf-binding but also for swath-cutting. The stalks then lie in swaths for a period of thirty to forty days, repeatedly getting wet from dew and then drying out again. This is similar to pre-retting, which represents a hybrid form of dew retting and water retting. The pre-retted stalks require less time for water retting. Presoaked stalks prepared in this manner are bound together into sheaves, placed in stacks, and pressed into bales with the PKV-1 Russian swath-loading binder. This method also allows the farmer to skip water retting and immediately process the stalks.

6.2.7 Loading

The bales in the field (or, rather, on the stubble) should be loaded and transported directly after being pressed. If it rains and the bales get wet, they should be taken apart. In this case the sheaves should be placed in stacks to dry, and then pressed into bales once again. This extra work can be avoided if the processing facility receives the entire load directly after baling. Stalks pressed while still wet will not be accepted by the processing facility.

The bales are loaded onto trucks directly from the field, with a front-loading tractor that can load 100 metric tons (110 short tons) of stalks in a single day, which is between ten and twelve truckloads.

6.2.8 Transport

Fiber-hemp bales are 2.4 to 3.0 meters (8–10 feet) long. In order to take full advantage of the available transport capacity, the bales are loaded onto trucks at a right angle to the road. In Germany, for example, the load width thereby just meets the permitted width specified in the German Road Traffic Act.

After the bales are transported, the stalks remaining in the field (hemp of lower quality weeded out by the reaping machine) are gathered and burned, if burning is permitted. This step is essential, partly as a preventive measure against the hemp borer and partly to aid in the subsequent soil preparation. The remaining stubble, if it is left unburned, does not readily break down because of its high fiber content, and this hinders both soil preparation and subsequent work with the seed drill.

6.3 Western European Technology

The harvesting methods practiced in western Europe, in contrast to eastern European methods, do not focus on the production of long fibers. That is because long-fiber production requires water retting, a practice abandoned in western Europe for both economic and ecological reasons.

As was previously mentioned, the sequence of harvesting events and the determination of the harvesting time are dependent upon production goals. These practices are not different from eastern European practices, and are therefore not described again here. Generally speaking, the harvesting technologies for hemp straw can be distinguished by whether the stalks are only mowed or are also chopped.

6.3.1 Mowing

Since hemp stalks have a tendency to wrap around rotating parts, the use of a rotation mower is ruled out. However, growers have had good experiences with bar mowers. A stalk separator should be attached to the end of the beam support to effectively separate the mowed plants from those plants that are still standing. An orderly orientation of the cut stalks is achieved and the path is cleared for the tractor. The stalk separator consists of a thin gliding blade.

In France, modified machines from the lucerne harvest have been adopted for use in the hemp fields. The machines use a mower equipped with a cutting beam to mow the stalks and then transport them through various rollers. The rollers crush the stalks, which are partially decorticated. It is important to keep the cutting apparatus sharp; otherwise, yield losses will result.

A preprocessor similar to the French lucerne equipment can be used to dry the stalks more quickly and to obtain a more homogeneous field retting. This preparation device crushes, breaks, or tears the harvested plants. The water can then better evaporate from the stalks, and the bacteria responsible for the retting process can better penetrate the cortex. The advantages of preprocessing the stalks are increased manageability for subsequent harvests and a shorter time required for field retting and drying. The promising results from France have been further supported by tests conducted in Holland, where the harvested plants were crushed between a large roller (35 centimeters or 14 inches in diameter) and five smaller rollers (12 centimeters or 5 inches in diameter). As with all rotating parts, these rollers must be protected from stalk parts that could wrap around them and jam the equipment.

The main difficulty in the processing of the hemp stalks is in the danger of over-retting in the field and the degradation of fiber quality that may result.

6.3.2 Chopping

In the Netherlands, the hemp crop is chopped instead of being mowed. Chopping results in simpler management of the hemp stalks in the field. Furthermore, the processing technology used in western Europe does not require the entire stalk length.

The Kemper cutter, manufactured by John Deere, consists of a row-independent cutter that follows a chaff cutter. This machine cuts the plants into pieces 50–60 centimeters (20–24 inches) long and ejects the stalks into the field, where they lie in disorder.

By chopping the hemp stalks, farmers avoid the need for special or heavily modified machines during harvesting. Conventional harvesting machinery can be used to turn and press the chopped hemp plants.

6.3.3 Drying

Both the mowing and the chopping procedures require field retting and subsequent drying before the plants can be gathered. The plants should be periodically turned to accelerate the drying process; this can be accomplished with a swath or other turning device. It is not necessary to modify the swath if the speed is adjusted for the volume of biomass. In this manner, anything that can wrap around the moving parts, particularly in the mowed plant stock, can be avoided. Due to the high risk of damage caused by stalks wrapping around moving parts, turners are generally used only on chopped stalks. Periodic employment of the swath or another turning device has a positive effect on the drying time similar to the effect produced by the preprocessing (i.e., crushing) of the hemp stalks. Crops containing plants with an abundance of leaves dry better after numerous turns or swaths.

A minimum of two to three turns is required, depending on weather conditions. Turning is absolutely necessary following reaping or trimming, and at least once before pressing.

6.3.4 Field Retting

Retting of the hemp straw is necessary in order to avoid significant fiber loss in the subsequent fiber-separation process. Since water retting is not applied in western Europe for economic and ecological reasons, field retting is the predominant method practiced. The hemp straw is sufficiently retted when the fibers easily separate from the stalk when the stalk is bent, normally after two to three weeks of field retting.

Before the hemp is pressed into bales and stored, the stock should dry to 15–20% water content (80–85% dry matter). Experience shows that hemp straw dries more quickly than straw from cereal crops, and that hemp suffers only minimal quality loss with longer retting periods. Even under less than optimal weather conditions (as during August and September in Germany), field retting and drying is therefore possible.

6.3.5 Pressing

After being dried, the hemp stalks should be compressed for transport and storage. Both round and rectangular bales are suitable for this purpose. The advantage of round bales is that they provide better ventilation since they are not as tightly compressed, facilitating further drying. The rectangular bales are easier to transport and store, due to their weight and form. If the hemp stalks were not preprocessed or bent, their length can cause a problem during pressing. Some form of preprocessing or bending is necessary in order to achieve tighter compression and better cohesiveness of the bales. Chopped hemp plants are ready for pressing, and require no further processing.

According to test results from the Netherlands, a rectangular bale press with a prepressing channel can accommodate both crushed and uncrushed hemp and provide stable bales. A rectangular bale press with a continuous pressing channel produces less stable bales that are also not entirely separated. This is mainly caused by individual stalks that are linking the bales. Round bales are the standard type produced in France.

When conventional presses are used, it is very important to avoid introducing anything that could get tangled in the equipment. The most significant problems occur when the straw is being loaded with the pickup device and during subsequent transport. Low speed alleviates some of these problems. This precaution is not always sufficient and sometimes the pickup device must be cleaned out on the field, which results in delays and cause significant time loss on larger acreages.

Due to the toughness of the hemp stalk, the process is hard on the equipment—especially cutting blades, counter blades, and pistons. These tools should be well-maintained to avoid any problems during pressing.

In the Netherlands, rectangular bales are pressed by self-moving presses. The bales weigh about 300 kilograms (660 pounds) and have a moisture content of 14–15% (the maximum moisture content is 18%). Pickup devices have proved to be a problem at this stage of harvesting, but the flow of stalks through the machinery proceeds with very few problems.

6.3.6 Silage

The difficulty with the previously described harvesting procedure is its dependence on weather conditions. In practice, the duration of the drying process is difficult to determine because of unpredictable weather conditions throughout most of Germany and western Europe in August and September. Therefore, a loss in quality as well as additional problems in meeting the sowing deadline for the following crop may result.

These observations have led to increased efforts to develop alternative harvesting techniques. The silage of straw is an option that has already been adopted in the harvesting of grass and corn. The advantage with this method is that the hemp stalks no longer dry in the field but can be directly loaded into the silo. The risk to the harvesting process posed by adverse weather conditions is then minimal. In addition, by eliminating the field-drying procedure, the work load is reduced and the harvesting deadline can be extended. This permits the cultivation of late-maturing varieties of hemp with higher dry-stock yields.

From 1990 to 1993, tests were conducted in the Netherlands to determine the best methods of harvesting wet plants. Hemp plants processed in this manner were to be used in the production of paper. However, application of this method is not yet advised because the effects on fiber quality and the subsequent utilization possibilities for those fibers have not yet been fully investigated.

6.4 Stalk Classification According to Industry Standards

Stalk quality is significantly influenced by the harvesting technique employed, and only slightly affected by the weather conditions during harvesting.

The amount and quality of fiber extracted from the stalks determines the certified quality and value of the stalks, particularly in the case of long fibers. Consequently, stalk values vary according to the standards of the fiber industry regarding fiber content. Determinations are made based on characteristics that are not externally visible or measurable. For the improvement of growing techniques, it is fundamentally important to strive toward growing plants with stalks that contain fine fibers and plentiful fibers. A reliable index for the content of fine fibers is the numerically expressed ratio of length and diameter of the stalk. Stalks that are smaller in diameter and longer produce a finer, more profitable fiber. The current growing technique in Hungary at harvest time prescribes a stand of 2.0–2.5 million plants per hectare (0.8–1.0 million plants per acre).

Under average growing conditions, the thickness in the midsection of the stalk is 5–7 millimeters (0.2–0.3 inches). If the stem's thickness does not exceed 10 millimeters (0.4 inches) and its length is at least 150 centimeters (5 feet, or 85% of the total plant length), it is classified as first grade according to industry standards.

Inferior quality results if the stalks age beyond the point of technical maturity as a result of late harvesting, or if the stalks turns grey to black as a result of rainfall. Grey discoloration of the stalk is evidence of fungal infection, which not only damages the epidermis but also affects the fibers. Fibers extracted from grey stalks lack tensile strength. If proper growing and harvesting methods are adopted, the stalk color will meet first-grade standards.

The SchSK 2.1 combined mower/binder places certain qualitative demands on the hemp plants. The machine separates shorter hemp plants, cuts off the plant portion below the soil, and organizes the stalks so that they lie parallel to one another and the stalk ends are uniform and level. The SchSK 2.1 then binds the stalks with natural sisal or hemp string. The string is wound around the center of the stalk and at the midsection of the stalk length with a measured diameter of 15–20 centimeters (6–8 inches).

If the harvesting machine is correctly adjusted, the sheaves will meet prescribed requirements. The bales pressed by the RKB-1 bale-press weigh 300–400 kilograms (650–900 pounds). They are suitable for loading on a tractor because each bale is tightly bound by a metal wire. If the machine is operated correctly and functioning properly, the bales will meet quality standards. New baling techniques have resulted in classification of the stalks into higher quality classes. The fiber-hemp stalks, bound into sheaves, are inspected and graded when the sheaves are handed over for sale. This usually takes place at the fiber-processing plant. Since it is not possible to conduct subsequent tests without breaking the bales apart, the quality standard is defined in the field during pressing. The current standards in Hungary are outlined as an example in Table 14. Such standards can be established in other countries according to the specific requirements of each individual country.

Please note: Since the German edition of this book was written in 1996, there have been several hemp harvesting machinery developments. As was mentioned previously, the Kemper cutter, which the Dutch company HempFlax worked to improve for hemp-harvesting usage, is now manufactured by John Deere. Hempline, Inc., of Ontario, Canada, which planted Canada's first modern hemp crop in 1994, has modified and improved several types of existing hemp-harvesting equipment.

Table 14: The Hungarian standard 17-631-74 relevant for the quality of raw fiber stalks

Quality Classifications	Grade I	Grade II	Grade III
Minimum technical length of 85% of the hemp stalks	150 cm (60 in)	100 cm (40 in)	60 cm (25 in)
Maximum thickness of 85% of the hemp stalks	10 mm (0.4 in)	12 mm (0.5 in)	14 mm (0.6 in)
Color	at least 80% yellowish light green, light brown	at least 60% light green, light brown (not important if technical length >150 cm)	not important
Maximum allowable impairments or damaged stalks in each bale	12%	25%	35%
Maximum moisture content	16%	16%	16%
Impurities (foliage, lesser-height plants, and brittle stalks)			
without correction	3%	4%	5%
including correction	12%	16%	25%

Chapter 7

■ ■ ■ ■ ■ ■ ■

Hempseed Cultivation

I n the regions where southern hemp varieties are traditionally cultivated (Hungary, Romania, Bulgaria, southern Russia, southern Ukraine, former Yogoslavia, and Turkey), seed-hemp cultivation differs from fiber-hemp cultivation with respect to growing areas and techniques.

In these countries, hemp grown for seeds is planted widely spaced, normally in marshy soils. Plants grown under such conditions produce significantly more seeds than those grown in dense stands. Hemp plants spaced farther apart develop more branches and grow taller (see Figure 24). Thus far there have been relatively few advances regarding mechanized methods of harvesting hemp seeds.

Figure 24: A widely spaced seed hemp crop (4 meters, or 13 feet high)

In Hungary, a solution to the need for mechanized reaping and threshing will soon be developed. Nevertheless, manual labor will still be necessary. In the other countries listed above, hemp seeds are still exclusively harvested by hand. In the traditional hemp-farming countries, hemp seeds serve only as seeds for the sowing of new crops. Seeds that sprout poorly are used for birdseed or are pressed for oil.

Because the highest seed yield that can be expected is one metric ton per hectare (900 pounds per acre), and because the oil content of hemp seeds is comparatively low (30–32%), hempseed oil is comparatively expensive. In spite of this, people in such western European countries as Austria, Switzerland, and Germany have expressed strong interest in seed-hemp cultivation because of the high nutritional value of the seeds and oil. Thus, development of a mechanized method of harvesting hemp seeds is imperative.

The German climate is not suited to profitable production of seed-hemp crops. The plains, which are amenable to hemp cultivation, are mainly situated in northern Germany. The landscape in central Germany is very hilly, with plateaus. In the south, where the climate would be suitable, the higher altitude is a negative factor. Nonetheless, there are smaller microclimate areas in Germany where the early-maturing varieties of hemp may be used exclusively to produce edible seeds or pressing oil, but not for sowing.

7.1 Soil and Climate Requirements

In contrast to fiber hemp, seed hemp produces good yields in marshy soil. Marshy soil's high content of organic materials and nitrogen stimulates growth, but germination is often delayed because of the soil's slow drying. When plants are being grown for seeds, it is not important if the fiber is coarse or if it tears easily. Water buildup, however, can significantly reduce crop yield. Plenty of warm weather and long fall seasons in these areas are beneficial to the maturation of hemp seeds.

Clearly, seed hemp thrives in mineral-rich soils just as well as fiber hemp does (see Section 5.1.4). We are mainly referring to southern locations, where seed-hemp plants mature five to six weeks later than fiber-hemp plants and where maturation and harvesting continue into the fall season. Seed hemp should not be grown at altitudes higher than 200–250 meters (650–800 feet) above sea level, because there is no guarantee that even early varieties will mature. The cultivation of southern dioecious varieties for seeds is not recommended in Germany because of small seed yields and the uncertainty of maturation.

In general, wine-growing regions are excellent areas for hempseed cultivation as well. Southern hemp, particularly when grown for seeds, requires a "wine-growing" climate. Wherever wine grapes grow, early- to mid-maturing monoecious and hybrid varieties of hemp can flourish.

Fiber and seed hemp's climatic requirements differ. The cultivation of hemp for seed production demands more heat because the vegetation period lasts five to six weeks longer. Thus the total heat quantity needed for early-maturing varieties is 2,300–2,500 GDD_C (growing degree days, 4,100–4,500 GDD_F).

7.2 Crop Rotation

It matters little what crops are grown prior to hemp; in fact, hemp can be grown on land where any other plant has been harvested. Hemp is a plant with "pioneer" characteristics: it thrives in cleared areas, plowed pastures, meadows, and drained bogs. Note that this is more a characteristic of seed hemp than of fiber hemp. When plowing pastures or meadows, it is very important to watch for potentially high nitrogen losses.

Hemp is a very resilient plant. Rotting problems are minimal even in monocultures. Nonetheless, it is impractical to grow hemp in the same location for more than two successive years, because in that case the hemp flea and particularly the hemp moth can rapidly multiply. Seed hemp is a valuable pre-crop, but it should not be planted prior to winter grains (even in highly fertile soil) because hemp seeds are harvested late in the season. However, it is excellent for summer crops because the field will be weed-free after the harvest. The same is true of fiber hemp (refer to Chapter 8).

7.3 Fertilization

Dual-purpose hemp crops, with an assumed seed yield of 0.6 metric tons per hectare (550 pounds per acre) and a straw yield of 4–6 metric tons per hectare (2–3 short tons per acre), extracts about 140 kilograms per hectare of nitrogen, 70 kilograms per hectare of phosphorous, and 120 kilograms per hectare of potassium (125, 60, and 105 pounds per acre) from the soil. Accordingly, organic or mineral fertilizer must be provided to the plants. Without organic fertilizer, the following quantities of mineral fertilizer are necessary for a standard seed yield and corresponding stalk yield:

▓ 400–500 kilograms per hectare (360–450 pounds per acre) of 25% calcium ammonium nitrate (100–125 kilograms N per hectare, or 90–110 pounds N per acre)

▓ 400–500 kilograms per hectare (360–450 pounds per acre) of 18% superphosphate (70–90 kilograms P_2O_5 per hectare, or 60–80 pounds P_2O_5 per acre)

▓ 300–400 kilograms per hectare (270–360 pounds per acre) of 40% potassium (120–160 kilograms K_2O per hectare, or 105–140 pounds K_2O per acre)

These amounts are suitable for an average yield, but the required amounts of P_2O_5 and K_2O could increase by 20–25%, depending on results from soil tests. In marshy soil, we highly advise increasing the amount of potassium by 50% and decreasing the amount of nitrogen. It is not necessary to increase nitrogen amounts to more than 100–120 kilograms per hectare (90–105 pounds per acre) because it may delay maturation. On mineral soils seed hemp tolerates higher doses of nitrogen as long as P_2O_5 dosage is also increased. Finally, as a general rule regarding minerals it is unjustified to use more than 100–120 kilograms per hectare (90–105 pounds per acre) of potassium, because mineral-rich soils are well supplied with this macroelement.

Results from soil tests should be compared to the limits for P_2O_5 and K_2O presented in Tables 9, 10, and 11. In Hungary, superphosphate is applied during plowing in the fall; potassium and nitrogen are later added in spring. Nitrogen can be applied once as a basic fertilizer in fall, at 50% in fall and 50% in spring, or it can all be applied in spring.

Organic fertilization is currently a hot topic among farmers. Marshy and mineral-rich soils provide excellent nutrients to seed hemp. The normal quantity of liquid manure with an average NPK content is 30–40 cubic meters per hectare (3,000–4,000 gallons per acre) in mineral-rich soil and 20–30 cubic meters per hectare (2,000–3,000 gallons per acre) in marshy soil. The liquid manure should be sprayed onto the soil prior to fall plowing. It is also necessary to introduce 30–50 kilograms of nitrogen per hectare (25–45 pounds per acre) in January before sowing, to balance the temporary nitrogen shortage. It is important to guard against nutrient washout when fertilizing, particularly in the fall. It can be minimized, for example, by working the fertilizer in with an intermediate crop.

7.4 Soil Preparation and Sowing

Soil preparation for both seed hemp and fiber hemp is identical. On marshy soil, however, preparation can occur two to three weeks later because the soil is wet and takes longer to dry. This initial delay is later compensated by the more rapid growth in the marshy soil.

■ *Sowing*

Since mechanical harvesting of seed hemp is not yet possible, row spacing cannot increase beyond the point where the ground is no longer overshadowed by plant foliage. The distance between each row can be 20, 30, or 40 centimeters (8, 12, or 16 inches). Increasing the distance between rows will increase the seed-bearing zone and plant height, which will in turn increase overall seed yield, particularly during periods of drought. Plants in rows separated by more than 40 centimeters (16 inches) will have larger stem diameters and more branches. In fact, these two factors may prohibit the use of a combine harvester to cut the plants. In addition, row separation of more than 40 centimeters (16 inches) means that there is no guarantee that the plants can overshadow the soil, and may require the use of a herbicide.

At a row spacing of 20 centimeters (8 inches), the required seeding rate is 50–60 kilograms per hectare, corresponding to 250–300 plants per square meter (45–55 pounds per acre and 20–25 plants per square foot) if the stalks are harvested for cellulose. The seeding rate for seed hemp, however, can be decreased by 50% to 25–30 kilograms per hectare (125–150 plants per square meter). If the distance between rows is increased to 30 centimeters (12 inches), the seeding rate decreases to

18–22 kilograms per hectare (90–110 plants per square meter). At the greatest row distance of 40 centimeters (16 inches), seed quantity is roughly 12–15 kilograms per hectare (60–75 plants per square meter). In all three instances, the number of seeds per linear meter of row is 20 (7 per foot) and the thousand grain weight is projected at 20 grams.

Single-use cultivation of seed hemp would require much greater row separation and fewer seeds than would the cultivation of fiber hemp, or dual-usage cultivation. In Hungary, hemp for seeds is sown with a precision seed drill and the rows are separated by 70 centimeters (28 inches), with 8–10 seeds per meter (2–3 seeds per foot). The seeding rate is 2–2.5 kilograms per hectare (1.8–2.2 pounds per acre). At this low density (10–15 plants per square meter), plants grow 3.5–4 meters (11–13 feet) high and develop numerous branches (see Figure 24). Correspondingly, the plants produce many seeds and are drought-resistant. It is impossible to use a combine on such a tall, outbranching crop. At such large distances between rows, crops may also need herbicide treatment, and they must be tilled with a cultivator.

The suggested shorter row separation discussed previously assures that the plants will overshadow the ground. Thus, herbicides will not be necessary. The same holds true for the French harvesting technology. Table 15 shows seeding rates for row separations ranging from 20–70 centimeters (8–28 inches) and 20 plants per linear meter (6 plants per linear foot).

Table 15: Required seed density for seed hemp as a function of row distance (individual plant distance = 5 centimeters, or 2 inches)

Row Width cm (in)	Seed Quantity kg/ha (lb/acre)	Sprouted Plants per ha (half acre)	Sprouted Plants per m³ (per ft²)
20 (8)	25–30 (22–27)	1,250,000–1,500,000 (500,000–600,000)	125–150 (12–14)
30 (12)	18–22 (16–20)	900,000–1,100,000 (360,000–450,000)	90–110 (8–10)
40 (16)	12–15 (11–13)	600,000–750,000 (240,000–300,000)	60–75 (6–7)
50 (20)	10–12 (9–11)	500,000–600,000 (200,000–240,000)	50–60 (5–6)
60* (24)	8–10 (7–9)	400,000–500,000 (160,000–200,000)	40–50 (4–5)
70* (28)	6–8 (5–7)	300,000–400,000 (120,000–160,000)	30–40 (3–4)

* These row widths can be sown with a precision seed drill at 2.0–2.5 kilograms per hectare (1.8–2.2 pounds per acre).

7.5 Seed Hemp

The use of herbicides is unnecessary if the rows are separated by 40 centimeters or less, because the hemp plants then suppress sprouting weeds. If the rows are separated by a distance greater than 40 centimeters, the weeds must be destroyed by chemical herbicides, by mechanical means with a cultivator, or by a combination thereof, since herbicides are generally effective for only four to six weeks. Should weeds develop after this period, they must be destroyed by mechanical means. At a row separation of more than 40 centimeters (16 inches), the plants cover the ground only after reaching a height of 150–200 centimeters (60–80 inches). The cultivator can be used as long as the plants are shorter than the grubber's height setting, or up to a plant height no greater than 60–70 centimeters (20–30 inches). Otherwise the grubber will damage the plants. If no herbicides are applied, the cultivator is not used until the plants have reached a height of 8–10 centimeters (3–4 inches) in order to avoid them being covered with soil.

If, for any reason, the rows are separated by more than 40 centimeters (16 inches), the use of herbicides may become necessary. In the following, we will discuss typical dosages for herbicides commonly applied in hempseed farming in Hungary.

Should it be necessary to suppress monocotyledons (weedy grasses), we recommend 8–10 liters per hectare (0.9–1.1 gallons per acre) of a solution containing 20% Benefin, incorporated into the soil with a disc or similar mechanism prior to sowing. Benefin provides complete protection against various *Setaria* and *Echinochloa* varieties.

One of the following solutions is recommended to fight dicotyledons, and should be introduced directly after sowing:

- 3–5 kilograms per hectare (2.7–4.5 pounds per acre) Maloran 50WP (Chlorobromuron 50%), *or*

- 1.5–2.5 kilograms per hectare (1.3–2.2 pounds per acre) Patoran 50WP (Metobromuron), *or*

- 5–6 kilograms per hectare (4.5–5.4 pounds per acre) Pyramine (Chloridason)

Maloran and Patoran are manufactured by the chemical firm CIBA-Geigy. These herbicides provide protection against weeds for a period ranging from six weeks to two months. The herbicides are effective only on mineral soils. On marshy soils, they are bound by organic matter and are not effective in the stated quantities. In greater quantities, they are toxic to the hemp plants.

Protection of seed hemp is necessary soon after the plants begin sprouting, particularly in the case of dry, warm weather, when the hemp flea could infest fields with such a low plant density. Refer to Section 5.6.2 (Insects and Other Pests) for a discussion on fighting the hemp flea.

Hemp borers damage fiber hemp and seed hemp, and their eradication is just as important as the eradication of the hemp flea. Stems attacked by the hemp borers swell up and develop knots. The stem becomes weak and can bend or snap easily. The damaged stem can break even during a relatively light wind which makes it difficult to harvest. The hemp borer also attacks the hemp seeds. A parathion solution should be applied immediately after the occurrence of borers, and we recommend spraying weekly thereafter. Usually the hemp borer infests only the outer perimeter of the field, so it is only necessary to spray these sections (an area 10–15 meters or 30–50 feet wide).

7.6 Harvesting

Determining when to harvest is difficult because it can be tricky. The seeds from different plants not only mature at different rates; they also mature at differing times on the same plant. When the lower seeds near the stalk are mature and have split open, the seeds near the top of the plant are not yet mature. The goal is to determine a harvest point at which minimal seed loss occurs.

Hemp seed can be harvested when the seed husk is hard in the seed-yielding section of the plant; the seeds' marble-like characteristic is easily identifiable and their external husks are yellow to bright-green in color. At this stage, the seeds do not fall off the plant if it is touched. Premature harvesting will result in numerous nonviable seeds, while a late harvest will result in significant yield reduction. This is particularly common in fields where the row distance is more than 40 centimeters

(16 inches). When the distance between rows is within the recommended 20–40 centimeters (8–16 inches), maturity is relatively uniform and the seeds mature, depending on plant variety, from mid-September to October 10. Seeds from the early-maturing monoecious varieties (Ferimon, Fedrina 19, Felina 34, Beniko, JUSO varieties, etc.) are mature by October 10 in central Germany.

When the seeds reach maturity, reaping should begin immediately. A combine harvester equipped with a dual-beam cutter can be employed, but should only be used during sunny, dry weather; that is, after the early morning mist has dissipated and when the sun is shining. Use of a combine harvester that can work with an elevated cutting table is also recommended (this is customary with the French harvesting technology for dual usage of seeds and fibers).

The seeds should be artificially dried at a maximum temperature of 40 C (105 F) if their moisture content exceeds 12%. The freshly threshed seeds have an average moisture content of 16–20%. This may require subsequent treatment if losses are to be avoided.

Cleaning seeds in the field is not very difficult, and can be done with a winnower. It is advisable to use a clod sieve with round slits of 5–6 millimeters (0.20–0.24 inches) and a seed sieve with elongated slots 1.5–2 millimeters (0.06–0.08 inches) in length. Depending on the grower's particular requirements, a seed separator could also come in handy. Smaller, nonviable seeds can easily be removed when sorting by density. It is difficult to remove the sclerotia from seeds that have been improperly treated; significant losses may also result when using a winnower or separator for that purpose.

The standard specifications for Hungarian hemp seed are outlined in Table 16.

As is the case with all oil seeds, special care is required in storage. Initially, the seeds can only be stored in thin layers and must be frequently turned. At a moisture content of 12%, seeds can be stored in sacks. The seeds cannot be stored in a dry storeroom because they will then no longer germinate. Hemp seeds burst easily, and can become rancid or germ-infested as a result.

Table 16: The Hungarian hempseed standard specifications
(MSz 6365-78) regarding hempseed quality

Quality class	Purity (minimum) %	Other seeds		Moisture content (minimum) %	Germination potential %
		Total	Weeds		
I*	99.5	20	5	12	90
II*	98.5	100	10	12	85

* The seeds cannot be contaminated with *Sclerotia* sp. and/or *Botrytis cinerea*

Recent experience suggests that loss of germination potential can be avoided if seeds are kept in cold storage. The temperature was kept between 0–5 C (32–40 F), and the humidity was set very low. Under these conditions, a germination rate of 80% can be maintained, even in the second year. Through storage at –10 C (14 F), the germination rate of the seeds after 5–6 years is 90%.

When farming hemp for seeds, the remaining stalks must be destroyed. There are still no environmentally sound methods of accomplishing this task. In Hungary the fields are burned, which works well when the stalks have a moisture content of 12–16%. Since field burning is prohibited in Germany, an alternative option is to chop the stems and stubble as finely as possible with a chaff cutter. However, experience with various equipment for this task is limited. The diameter of the stems should certainly not exceed 10–15 millimeters (0.4–0.6 inches). After plowing, roughly 50% more nitrogen should be added for the next crop. The additional nitrogen aids in the decomposition of the wood that remains in the soil.

7.7 Seed Harvesting and Dual Usage in Western Europe

When hemp is grown for seeds in western Europe, the plant is predominantly cultivated for dual usage as seed and fiber. Seed quantities ranging from 20–50 kilograms per hectare (15–45 pounds per acre) are used. With such seed quantities, the plant density and degree of lignification allow for the use of conventional combine harvesters.

In France, hemp is grown only for its seeds in a relatively small area. When growing hemp in this manner, it is necessary to cut the plants to harvest the seeds. Threshing with a combine harvester can take place after the plants have dried in the field for two to three days. Due to the considerable strain on the harvester, specifically modified combine harvesters are used in which straw and seeds pass through modified rollers, avoiding damage to bearings and the wrapping of fiber around any rotating parts. The seeds' moisture content should not exceed 9–10% at harvest. If the seeds are to be stored, their moisture content should be 8% or less.

The harvesting of the plants for both seeds and fiber is a two-stage process. In the first stage, the tops of the plants are cut off and threshed. The cutter from a modified combine harvester is set at the lower end of the seed-bearing sections of the plant. The combine harvester must maintain a high speed to insure a trouble-free harvest. Only then will the cut plant pieces fall directly into the chopper, which should be narrowed to avoid blockages. An unmodified combine harvester could be used, but the type of machine and the driver's skills are critical factors to consider. Axial-flow combines have proven to be well suited for this type of harvesting. The second stage involves harvesting of the stalks remaining in the field. This step is essentially identical to the harvesting process outlined in our previous discussion on fiber-hemp cultivation. The stems are cut, possibly trimmed, and pressed into bales after retting and drying.

Chapter 8

■ ■ ■ ■ ■ ■ ■ ■

An Ecological Evaluation
of Hemp Cultivation

Since Germany's rediscovery of hemp in 1993, the terms "hemp" and "ecology" have become more closely associated in the German language. Hemp is now regarded as *the* environmentally friendly natural resource. According to one report, "Hemp is well suited for sustainable, ecologically sound agriculture, which is in increasing demand in Europe. Hemp—almost automatically—provides clean bioresources that are ready for the development of environmentally friendly products." (Hingst/Mackwitz 1996)

In fact, new "organically certified" hemp product lines have been developed, primarily by eco-companies in western Europe. These companies demand raw materials that are clean as well as environmentally friendly, and they incorporate the "eco-hemp" image in their marketing campaigns. The hemp leaf has become a symbol for the restructuring of the market into an ecological, regional, closed-cycle economy.

Those readers who have a general understanding of hemp may refute some points; for example, points outlined in the detailed discussion on plant protection and chemical defoliation. One may ask, "Isn't hemp a plant that is automatially cultivated using environmentally sound methods and then processed as is?" The answer is: of course not!

Chemical Usage in Traditional Hemp Cultivation

To the authors, environmental awareness means hiding nothing. It means not making positive claims just because they sound good, but instead clarifying the facts. This book reports where and why the use of chemicals has been and remains to this day customary, e.g., in Hungary and other eastern European countries where hemp has been cultivated over the last few decades. The truth is that, in these countries, hemp, just like other crops, has been cultivated by conventional methods. It was the western demand for "eco-hemp" that first sparked an interest in organic cultivation in eastern Europe.

This book shows that, in comparison to other crops, even conventionally cultivated hemp grows quite well with very few chemicals. Therefore, in comparison to most other plants, it is easier to grow hemp without any chemicals. The experiences of both old and new hemp-growing countries in western Europe verify this. Hemp crops are not treated in France, England, Germany, or the Netherlands. There is simply no economic necessity that justifies the use of chemicals. Pests, diseases, and weeds cause no significant reduction in crop yield. Furthermore, uncontaminated raw materials yield premium market prices. In western Europe, the notion of hemp as an ecological crop has an element of truth.

An Ecological Assessment of Hemp Cultivation

The purpose of this chapter is to critically examine the life of hemp —from the time it is a seed to its maturation as fiber—in order to determine its ecological value and to provide some recommendations.

▨ Weed Control

Following the establishment of the plant stand, one of the most important ecological advantages of hemp soon becomes evident. Due to hemp's rapid early growth and the density of the crop, strong weed suppression is virtually guaranteed. Even thistles and couch grass are killed off by hemp.

This extremely effective weed suppression is only found under two conditions. First, the hemp plants must reach a decent developmental stage, which primarily requires a sufficient supply of water and nutrients, good soil structure, and the avoidance of moisture buildup and soil compression. Cultivated in poor soil, hemp stands no chance against weeds.

Secondly, weed suppression can only be achieved with fiber hemp crops that are sown in relatively high seed densities. If hemp is sown with much less than about 40 kg per hectare (36 lbs per acre), the crop is unable to sufficiently expand and weeds have a greater chance of surviving. This is especially a problem in the cultivation of hemp for seeds. Ideally this problem should be solved ecologically by mechanical methods, e.g., the use of hoes and harrows. The wide plant spacing makes this task somewhat easier.[1]

Fiber hemp restrains nearly 100% of weeds. From an environmental point of view, this results in the following:

■ Under normal conditions, the cultivation of fiber hemp requires no herbicides. This is clearly a significant ecological advantage in comparison to most other crops.

■ In addition, hemp is extremely sensitive to most herbicides. Even herbicide residuals from preceding years cause hemp to have a weaker growth rate. Those who want to cultivate hemp should use herbicides with caution during the entire crop rotation.

■ Hemp restrains weeds so well that they rarely mature. This weed control carries over to the following crop, which is of particular interest to organic farmers, who are always combating the growth and spread of weeds.

■ *Eutrophication and Soil Erosion*

In "Hemp Product Lines From an Ecological Viewpoint," a part of the Hemp Product Line Project (nova 1996), the ifeu Institute writes: "The potential danger of soil erosion by various plants is chosen as the criterion for an index describing eutrophication of lakes and streams as a result of agriculture." Hemp does favorably here in two respects. For one, because of its high yield per acre, hemp farming needs a relatively small area. Secondly, about three weeks after seedling emergence (at the middle or end of May), hemp provides complete ground cover,

---

[1] It is particularly important to avoid the use of herbicides (or other pesticides) with hemp for seeds because certified organic cultivation (KbA) is important with the nutritionally/physiologically high-quality seeds. The market prices for seeds vary between conventional and KbA-products by a factor of 2–3.

and three weeks after harvesting in early September, the leftover stalks from field retting effectively prevents soil erosion. Immediately afterward, winter cereal crops are sown. Thus, the fields are susceptible to only minimal erosion during the following winter. The danger of eutrophication from washed out salts, fertilizers, and soil particles declines in the following sequence:

Cotton > Soy >> Evening Primrose > Canola = Hemp

(Note: The above-mentioned plants were assumed to be substitutable by hemp in the ifeu Institute's comparative ecobalance.)

■ Soil Structure

Hemp's high degree of ground coverage and deep root penetration into the (noncompressed) soil has further environmental advantages. The plant provides excellent shade, soil quality is improved, soil ventilation and water balance are improved, and soil fauna thrives in the area. The 10–20% yield increase in winter wheat being increasingly reported by farmers who plant wheat directly after hemp is primarily due to improved soil structure and weed suppression. The company Hempflax, based in the Netherlands, reports that hemp fields dry off better and warm up more rapidly early in the year than do other fields, resulting in the possibility of earlier cultivation of other crops. Improved soil ventilation also plays an important role in the earlier cultivation of crops.

■ Fertilization

Organic farmers have discovered that hemp's nutritional requirements are quite high. Hemp requires a large amount of nitrogen fertilizer—about 80 to 120 kg N per hectare (70 to 110 lb N/acre). In organic farming, hemp and wheat compete for sites with excellent soils. For conventional farmers, hemp occupies a median position in relation to other crops. The German Environmental Protection Agency (UBA) recommends that no more than 120 kg N per hectare (110 lb N/acre) be used, which is quite sufficient for hemp crops.

An important factor for environmental assessments is the possibility of nitrogen leaching into bodies of water. Due to its long roots (up to two meters or 6.5 feet), hemp can extract residual nutrients from previous crops deep in the soil, thereby minimizing the danger of nitrogen leaching.

Because the gradual release of nitrogen into the soil is ideal for hemp, the crop is well suited for organic fertilizers such as liquid manure and dung from stables.

An abundance of organic matter stays in the field after hemp is cultivated: leaves, roots, and part of the hurds remain after field retting and provide nutrients to the subsequent crop.

In spite of hemp's need for fertilizer, organic farmers will have no problems if hemp is accurately placed in crop rotation. We advise planting hemp after clovers or a legume.

Besides nitrogen, an adequate supply of potassium is of great importance (refer to Section 5.2).

■ **Plant Protection** *(refer also to Section 5.6)*

Hemp is a plant that contracts few diseases and has few pests. Even with conventional cultivation methods, it flourishes with very little protection or use of pesticides. The minimal threat posed by insects is evidenced by hemp's high self-tolerance (self-compatibility in subsequent crops). One of the reasons for hemp's self-tolerance is the absence of any closely related crops cultivated in close proximity that would have the same pests as hemp.

The high self-tolerance of hemp could encourage the plant's cultivation in monocultures. Therefore, the following warning should be taken seriously: nutrient and pest problems must be expected when hemp is continuously cultivated in monoculture. (These problems, however, are not discussed in this book). But since hemp is an excellent addition to a crop rotation, and since its total acreage in Germany is not expected to exceed 40,000 hectares (100,000 acres), cultivating hemp in monoculture can easily be avoided. The high benefits of hemp as a preceding crop are not optimally utilized in monoculture.

Gutberlet and Karus summarize the findings from their study "Parasitic Diseases and Pests That Threaten Hemp" as follows:

> As expected, a multitude of parasitic diseases and pests exist
> that infest hemp plants. However, in contrast to many other
> crops, especially other renewable resources such as canola or flax,

economic loss due to infestation of hemp by pests and diseases is minimal. Most of the hemp pests are nonspecific and, if nearby, simply prefer other crops. There are practically no specific pests that cannot be controlled by technical cultivation methods such as seed purification or biological pest control. As a general rule of thumb, fiber hemp can be cultivated without the use of pesticides, and there is no danger of substantial crop loss.

In the study referred to above, the ifeu Institute writes:

Aside from the positive quantitative aspects of the essentially biocide-free hemp cultivation, qualitative aspects of the pesticides in use are important for the assessment of toxicological environmental influences. Cotton and soy, the two hemp substitute crops, are characterized not only by their enormous demand for pesticides, but also by the toxicity of the respective pesticides in use. The chief groups of active substances are organophosphates and synthetic pyrethroids. In contrast to the fungicides (dicarboximides) used during hemp cultivation, these groups have a high aqueous toxicity and represent a serious threat to all living creatures as well as the environment. . . . Based on these toxicological considerations, the following ranking can be derived for industrial crops:

Cotton > Soy >> Evening Primrose = Canola > Hemp

[Note: The above-mentioned plants were assumed to be substitutable by hemp in the ifeu Institute's comparative ecobalance.]

Consequently, there are important advantages for hemp textiles, even considering toxicological facts.

The knowledge and experience gained from the western European hemp-growing countries such as France, England, Germany, and the Netherlands prove that commercial hemp cultivation without the use of pesticides is in fact possible. Costs for pesticide use are out of proportion compared to possible crop losses. Studies conducted in the Netherlands showed that the preventive use of fungicides is generally not worthwhile, since gray mold infestation becomes a problem only if the early summer months are rainy. And if, by chance, the crop yield actually declines because no pesticides are used, the higher market price of the organically cultivated crop will more than offset the lower crop yield. In addition, the use of pesticides is now precluded by the present marketing of hemp's "eco-image."

Nevertheless, as soon as hemp is recognized beyond the environmentally oriented industries as an interesting resource, conventional farmers may choose to apply pesticides to their crops every few years— a routine still customary in eastern Europe. Presently, the German pest control agencies do not recommend the use of pesticides on hemp. These recommendations will likely be reconsidered as hemp acreage continues to expand.

In order to prove that hemp is not sprayed in conventional cultivation in eastern Europe and China, hemp fabrics and textiles are subjected to frequent tests to detect the presence of harmful substances. In fact, analyses by order of the nova Institute show that all hemp fabrics from Hungary, Romania, and China that have been tested were free of harmful substances and in accordance with standards established by the Natural Textile Association (Arbeitskreis Naturtextilien, AKN).[2]

What conclusions can be drawn from the standards established by the AKN? First of all, consumers who purchase hemp textiles undoubtedly profit. They can be relatively certain that the products are pesticide-free, even if this is not explicitly stated on the merchandise. This holds true even though the present AKN standards are tailored to cotton, wool, and flax. But, as it is, most pesticides presently used on hemp

---

[2] This is true except for one case. The first hemp textiles from Hungary, which were available on the German market and tested in 1994, showed high residues of PCP (Pentachlorophenol). An investigation in Hungary and Romania revealed that PCP was not used during cultivation but rather during transport of the processed long fibers from Romania to Hungary. In Romania it was customary to spray all natural fibers with PCP before transporting them, in order to prevent a possible mold infestation during transport and loading. Due to pressure from German businesses, this practice was immediately discontinued without other problems emerging. This was confirmed by later analyses.

were developed for use on cotton. Therefore, these pesticides are already included in the AKN standards. It remains to be seen whether other pesticides currently not included on the AKN list will be used on hemp.

Another problem is the detection of pesticides used during cultivation in processed fibers and textiles. Even with cotton, it is very difficult to prove that residuals from pesticides have been passed along to processed fibers or fabrics. Often, only those pesticides used on plants shortly before harvesting (for example, for defoliation) could be detected when the cotton pods were already open. This is all the more difficult to determine with hemp, because the fibers are well protected by the bark throughout the entire vegetation period. Even systemic pesticides penetrate fibers very little because the fibers serve as a support system rather than a transport system.

Conclusion: As with cotton, one cannot conclude from residue-free textile fibers that cultivation has been pesticide-free. On-the-spot controls continue to be effective, in fact, irreplaceable.

■ *Characteristics of Hemp in Crop Rotation*

Hemp is an excellent plant to use in a crop rotation. It is a neutral crop that excellently conditions the soil for the subsequent crop. Hemp does react severely to herbicide residuals, and in their presence crop problems can easily result. It makes an outstanding preceding crop, particularly since it leaves behind an exceptional soil structure (see above). Fiber hemp virtually eradicates weeds for the next crop. As was previously noted, farmers in increasing numbers are reporting crop-yield increases as a result of the use of hemp as a preceding crop. Also, hemp is perfectly suitable as an intermediate crop because it is sown at the end of April or beginning of May.

Tests in the Netherlands have shown that some varieties of hemp restrain certain types of nematodes. This discovery could become particularly important for the cultivation and crop rotation of potatoes.

■ *Harvesting and Field Retting*

In western Europe, hemp is mowed, cut to stalk lengths of 50–60 centimeters (20–25 inches), and remains in the field two to three weeks for retting. The chemical defoliation methods customarily used in eastern Europe were not adopted in western Europe. From a quantitative and economic point of view, chemical defoliation makes sense only if the straw is to be water-retted, a method ruled out in western Europe.

Field retting is economically advantageous. First, it hinders soil erosion (see above). Secondly, field retting provides the soil with organic matter and nutrients. The remaining stubble in the soil is also advantageous, because it provides yet another level of erosion protection. The stubble is not plowed under until shortly before the sowing of the next crop.

The late harvest period in September required by EU-subsidy regulations is ecologically problematic (refer to Section 9.4). The use of heavy harvesting machinery in September (a predominantly wet month) leads to the compression of the soil and the formation of deep grooves in the field.

Overall, the ecological significance of hemp cultivation is very positive. This viewpoint is supported by a study conducted in the Netherlands (Regional Development Group 1982). The study, considering ecological factors, investigated which new varieties of crops are particularly suited for cultivation in the Netherlands. The results favored hemp as a new (old) agricultural plant, and contributed significantly to the establishment of the Dutch Hemp Research Program.

In conclusion, it is worth citing the results of the ifeu Institute's study once more. In the study, the effects of the cultivation and harvesting of one ton of fiber hemp were compared to the same effects from one ton of cotton fiber.

The substitution of cotton with hemp in the textile industry would lead to considerable ecological advantages during the cultivation and harvesting stages. The reduction in primary energy consumption and emissions would be about three times per each ton of fiber (see Table 17).

More extensive analyses show, however, that—for lack of better processing equipment—hemp is presently forfeiting a significant portion of these advantages during fiber separation and processing (spinning and weaving).

Table 17: Ecological balance (primary energy requirements and emissions) for the cultivation and harvesting of hemp and cotton (from the ifeu Institute in nova 1996)

Reference quantities: One metric ton of fiber (one short ton of fiber)

Crop	PE GJ (10⁶ BTU)	CO₂ kg (lb)	N₂O kg (lb)	CO₂-Eq kg (lb)	SO₂ kg (lb)	NO₂ kg (lb)	SO₂-Eq kg (lb)
Hemp	8.2 (7.1)	544 (1,090)	1.3 (2.5)	947 (1,900)	1.2 (2.4)	4.5 (9.1)	4.4 (8.8)
Cotton	25.2 (21.7)	1,680 (3,370)	3.0 (6.1)	2,650 (5,310)	2.5 (5.0)	14.8 (29.7)	12.9 (25.9)

PE: Primary Energy Consumption

CO₂-Eq: CO₂-Equivalent, measure for global greenhouse potential

SO₂-Eq: SO₂-Equivalent, measure for total acidification

Chapter 9

■ ■ ■ ■ ■ ■ ■

New Uses for Hemp
in Western Europe

9.1 Economic Considerations of Cultivation and Harvesting

Traditional methods of harvesting and processing hemp, which are still being practiced in eastern Europe and Asia (refer to Sections 6.1 and 6.2), are essentially based on mowing the plants and placing them in parallel rows; drying the stalks on the field, followed by water retting; drying the retted straw; and finally breaking, scutching, and hackling the stalks to extract the long fibers. Secondary products derived from this process include tow fibers and hurds.

While these traditional methods continue to be successful in eastern Europe due to the available infrastructure, machinery, and experience, they have no such likelihood of success in Germany and other western countries because of the high costs, the limited production of adequate fibers, and the environmental problems (retting water) associated with the traditional methods. The high costs are primarily a result of the labor-intensive processes and the use of expensive modified machinery.

The modern hemp industry in Germany favors the modern whole- or short-fiber technologies, which means harvesting the entire stalk including the hurds, which contain short fibers. This technique has been successfully practiced in England and the Netherlands and was successfully employed during the first hemp-harvesting year in Germany (1996). The technological and environmental characteristics of this technique are being briefly introduced here (refer also to Section 6.3).

As soon as the fibers have reached technical maturity,[1] the hemp plants are mowed or chopped into pieces 50–60 centimeters (18–24 inches) long and deposited on the field for retting.[2] Field retting usually takes two to three weeks but, depending on weather conditions, may require more or less time. The harvested plants are repeatedly turned so that the stalks air out more uniformly. The plants are adequately retted when the fibers separate easily from the stalk by hand and the stalks have sufficiently dried (to a maximum moisture content of 15–18%). At this point the stalks can be pressed into round or rectangular bales and loaded for transport.[3] The first stage of mechanical fiber processing then follows, in which the stalks are broken apart and the fibers are separated from the hurds (compare with Section 9.2).

Table 18 outlines the results from detailed economic analyses of cultivation and harvesting costs conducted by the nova Institute and summarized in its Hemp Product Line Project report (nova 1996). The total-cost calculation accounts for all costs associated with cultivation and harvesting. Included in this calculation were costs according to the 1994 KTBL (German Committee for Agricultural Engineering and Construction Methods), as well as estimated costs associated with necessary

---

[1] The actual time of harvesting in western Europe is determined to a greater extent by EU-subsidy regulations than by the technical maturity (compare to Section 9.4).

[2] The respective harvesting technique is available and was already tested in the field (compare to Section 6.3); further optimizations are possible and already under development. Harvesting for seeds is more problematic: available reaper/threshers have difficulties with processing the plants. With dual usage, the upper one-third of the plant is cut off and threshed. The remainder of the plant is mowed or chopped. Presently this is still accomplished in two separate steps. Further modifications and new developments are already under development. (Lohmeyer 1996)

[3] Besides the field retting method, there are other alternatives that are still in the experimental stage. One example is the silage method, where the short, chopped segments of the hemp stalk are put in silage and the fibers are processed (compare to Section 6.3.6). A second example is the use of green hemp, in which case retting is foregone. This method is no longer recommended because the fiber loss during processing is greater and the green hemp fibers exhibit poorer qualities in most subsequent applications in comparison with retted fibers.

Table 18: Economic balance of total costs of hemp stalk supply (cultivation and harvesting) and estimated subsidy contribution at various yield levels*

	Hemp Stalk Yield – DM/ha (U.S. $/acre)**		
	6t/ha (2.7 short tons/acre)	8t/ha (4.0 short tons/acre)	12t/ha (5.4 short tons/acre)
TOTAL COSTS			
Cultivation and harvesting costs	−2,250 (−520)	−2,250 (−520)	−2,250 (−520)
Market potential for hemp stalks	840 (190)	1,260 (290)	1,680 (390)
EU subsidies 1996	1,510 (350)	1,510 (350)	1,510 (350)
PROFIT/LOSS			
Without EU subsidies	−1,410 (−330)	−990 (−230)	−570 (−130)
Including EU subsidies	100 (23)	520 (120)	940 (220)
RETURN ABOVE OPERATING COSTS			
Without EU subsidies	−490 (−110)	−70 (−20)	350 (80)
Including EU subsidies	1,020 (240)	1,440 (330)	1,860 (430)

* A market price of DM 140 per metric ton (U.S. $73 per short ton) for the cut stalks is assumed (chopped length = 50–60 centimeters, 20–24 inches)

** At an exchange rate of 1 DM = 0.57 U.S. $ nova 1996

modifications to machinery. The estimated return above operating costs was derived from total costs. Surplus turnover (market performance) was calculated over the variable costs.

The standard market price for chopped hemp stalk was assumed at DM 140 per metric ton (U.S. $73 per short ton), which was derived from recent interviews with British and Dutch companies. Even German companies, e.g., BaFa (Baden Natural Fiber Processing), aim at such standard market prices for chopped hemp stalk. The European Union subsidy was assumed at the 1996 value of DM 1,510 per hectare (U.S. $350 per acre) (compare Section 9.4). Transport distances of about 30 kilometers (20 miles) between the field and the processing plant were included in the calculations.

Based on current conditions (EU-registered varieties, cultivation and harvesting methods) average hemp stalk yields of 6–9 metric tons per hectare (3–4 short tons per acre) can be expected. In Table 18, the projected yield of 12 metric tons per hectare (5.5 short tons per acre) applies to varieties optimized for particular regions that produce yields higher than normal. The EU registration of such varieties can be expected in the near future.

Lesser harvesting costs are accrued if the plants are chopped and not scutched. However, since the market value of uncut hemp stalks decreases due to poorer manageability during processing, the total costs and return above operating cost estimates cited above remain largely unchanged.

By and large, only EU subsidies make hemp cultivation a profitable venture under current economic conditions. The point where fiber costs and market prices meet dictates which subsidies are necessary to cover current costs. Assuming a dry-stalk yield of 9 metric tons per hectare (4 short tons per acre) or a dual-usage fiber yield of 6 metric tons per hectare (2.7 short tons per acre), a minimum subsidy of DM 800–1,200 per hectare (U.S. $190–$280) would be required. The agricultural "set-aside" recompensation is hardly high enough to render hemp cultivation profitable. However, cost reductions in hemp-processing techniques can be expected in the near future. Higher crop yields from optimized varieties combined with lower harvesting costs due to new harvesting machinery (i.e., production of fiber and seeds in a single harvest) will contribute to the future feasibility of hemp cultivation. Should subsidies be reduced, higher market prices for fiber hemp will contribute to hemp cultivation's continued viability. Once fiber-processing factories and product lines are established, the processors will be able to pay prices for stalks ranging from DM 180–200 (U.S. $100–$110), because stalk prices only partially determine the fiber prices.

Some estimated costs for seed production will now be quoted. An estimated seed yield of about 1,200 kilogram per hectare (1,100 pounds per acre) is assumed when growing hemp for seed. In order to cover both cultivation and harvesting prices in the total cost estimate, a sale price of at least DM 0.20 per kilogram including EU subsidies (U.S. $0.05 per pound) or DM 1.50 per kilogram without EU subsidies (U.S. $0.40

per pound) must be achieved. With a projected market price of DM 2.00 per kilogram or U.S. $0.50 per pound (for certified organic quality the current market price is more than DM 4.00 per kilogram or U.S. $1.00 per pound) the total cost estimate would equate to profits of DM 2,140 per hectare (U.S. $500 per acre) including EU subsidies or DM 630 per hectare (U.S. $150 per acre) without EU subsidies. These estimates were made under the assumption that a modified harvester can be used to harvest the seed hemp, which presently is not the case but is expected in the near future.

Dual usage of fiber and seeds leads to slightly higher, equal, or probably even lower fiber costs, depending on seed yield and profitability. In the case of a late harvest, fiber quality issues may result from excessive retting or lignification. The reduction in quality can in turn reduce the market price for the hemp stalks. Dual usage therefore only makes sense with sufficiently high seed yields and market prices.

While the seeds can be integrated into the existing crop processing lines (e.g., oil extraction), the hemp fibers and hurds must be separated in a specialized fiber-processing plant. Separating the fiber into different groups (long fiber and tow fiber) by quality, which is the traditional method, does not occur with dual usage stalks. The entire fiber is produced as short fiber (whole fiber or short fiber use).

9.2 The Technology and Economics of New Fiber-Processing Methods

Depending on the desired product lines (compare with Section 9.3), either the most cost-effective mechanical means of fiber processing or a chemical-physical method of fiber processing is favored.

▪ *Cost-Effective Mechanical Processing of Coarse to Medium-Fine Fiber Bundles*

After drying, the field-retted hemp stalks are pressed into bales and loaded, and the hurds (the woody core of the stalk) are mechanically decorticated in short-fiber processing plants. The fibers are cleaned and the fiber bundles are refined to a pre- or mid-level grade. Adequate mechanical fiber-processing equipment or components will eventually be offered by mechanical engineering firms in Germany (Gebr. Bahmer

Söhnstetten, Temafa Bergisch Gladbach), Belgium (Charle & Co. Kortrijk-Bissegem), and France (La Roche). The new German fiber-processing equipment facilities were built under a government program for advancements in flax processing, whereas the other equipment is modified machinery from flax and hemp tow processing.

■ Process Sequence

Depending on the preceding harvesting method, the hemp stalks can be either broken, chopped, or partially decorticated. The stalks are provided in bale form to the processing plants. After passing through the bale opener, the stalks are directed through a dispensing or buffing device, which breaks the hurds and separates the fiber from the hurds. Depending on the intensity of the process, which is accomplished with various serrated rollers, a certain amount of loose hurds falls on the collecting belt that can later be supplied for client-specific uses. Rock crushers or metal detectors are occasionally used before the decorticator. The stalks that have been partially decorticated are then opened, cleaned, and broken apart. These steps represent the first stage of the fiber-processing process. Most hurds are then removed, and the coarse fiber bundles are cut and partially separated. Depending on the model, manufacturer, and machine design of the equipment used, the fibers may be further processed, refined, and cleaned. Stage cleaners, rollers, cards, pressure cleaners, or even pneumatic cleaners can be used. According to the desired degree of fineness or "cottonization" (the processing of hemp fibers to achieve qualities similar to those of cotton), the number and intensity of the processing steps may vary. The fibers are subjected to an extremely high degree of stress as they proceed through the various processing machines. Depending on the raw materials used (degree of maturation, degree of retting, etc.), fiber loss and fiber yield can vary. Mechanical "cottonization" always results in some damage to the fibers.

Mechanical processing turns out to be more involved with hemp than with flax. In addition, hemp's fiber yield and fineness are inferior to flax. This is even more so if hemp plants are field-retted rather than water-retted (water retting being the more traditional of the two methods). This is because the hemp fibers, in addition to their chemical bonds, are also mechanically connected. These mechanical connections develop because individual fibers grow from one fiber bundle into another. The interconnected fibers result in bonds that make it virtually impossible to completely separate the fiber bundles. Furthermore, the lignified cortex is extremely resistant to the retting process, which makes the complete penetration of the retting bacteria into the plant's cortex much slower than with flax.

While fiber mixtures for apparel with a fineness of Nm 20 to Nm 40 can be produced from mechanically processed flax fibers, mechanically processed hemp only reaches a level of Nm 5 to Nm 10 (field-retted) or Nm 10 to Nm 20 (water-retted). This lower fineness excludes hemp fibers from a significant sector of the apparel market. Since the new harvesting techniques employ field retting only, the prospects for mechanically processed hemp fibers for clothing are considered to be very minimal (see below, also refer to Section 9.3).

■ The Economics of Mechanical Fiber Processing

The price of mechanically produced fiber—assuming a hemp stalk price of DM 140 per metric ton (U.S. $75 per short ton), which currently is only feasible with the aid of EU subsidies (see above)—ranges from DM 0.70–1.10 per kilogram (U.S. $0.20–$0.30 per pound) for well-decorticated fibers to DM 0.90–1.40 per kilogram (U.S. $0.25–$0.35 per pound) for preprocessed/coarse-processed fiber, and DM 1.10–2.00 per kilogram (U.S. $0.25–$0.50 per pound) for medium-processed fiber.[4] Table 19 gives an overview of estimated prices for fibers of different qualities, depending on the number of work shifts.

Due to the higher yield per acre, hemp prices are below the prices of flax of a similar quality and are undoubtedly competitive with imported natural fibers.

---

[4] The fiber prices include a high-priced sale of the hurds from fiber processing (DM 200–300 per metric ton, U.S. $100–$155 per short ton). These prices can be achieved if the hurds are sold for animal bedding in which case they require careful cleaning and sorting.

Favored areas of use for the various fiber qualities are (with increasing degree of demand or refinement):

- cellulose (subsequent chemical breakdown)
- press-molded interior panels for automobiles
- geotextiles
- needle-punched carpeting
- fiber-reinforced synthetic materials

Under current conditions (i.e., without environmental benefits leading to economic advantages) the quoted prices are necessary to win and retain markets. In Europe this is presently only feasible with the aid of EU subsidies (compare to Section 9.4).

An increase in the price of hemp stalks, resulting (for example) from a reduction or discontinuation of EU subsidies, immediately and severely impacts the price of reasonably priced coarse to medium-fine hemp fibers. In order to retain the lower-priced markets that are already developed, successively higher harvest yields must be achieved. In addition, harvesting and processing methods must be technically and economically improved. Current EU subsidies offer a favorable basis from which to progress.

Another problem arises from product lines for which price is the main factor and in which fiber quality and even hurd content play a minor role. Press-molded interior panels for the automobile industry, for example, utilize low-cost coarse fibers. Good jute harvests in India can therefore threaten sales of German hemp fibers to the German automobile industry—at least, as long as environmental considerations such as transport distance or the use of herbicides and pesticides are not economically considered. Thus, the higher-priced fiber markets with more flexibility for price adjustments are absolutely necessary to support the low-priced fiber markets.

Higher-priced fiber markets are very difficult to develop and retain with mechanically processed hemp. In contrast to flax, the hemp fibers require further mechanical refinements if they are to be sold in the textile industry or insulating-materials markets; such refinements appear to be technically and economically unfeasible for field-retted hemp. Experience shows that hemp fibers require water retting before

Table 19: Estimated minimum sale price for fibers of various qualities depending on number of work shifts (prices at the processing plant, including high-priced sale of the hurds)

Process Technology, Fiber Quality (Company)	DM/kg fiber (U.S. $/lb fiber)*		
	3 work shifts	2 work shifts	1 work shift
1) Mechanically well decorticated, fine fibers (Charle)	0.71 (0.18)	0.80 (0.21)	1.09 (0.28)
2) Mechanically processed, coarse to medium-fine fibers (Charle and La Roche)	0.95 (0.25)	1.07 (0.28)	1.42 (0.37)
3) Mechanically processed, coarse to medium-fine fibers (Temafa)	0.89 (0.23)	1.02 (0.26)	1.40 (0.36)
4) Mechanically processed, fine fibers (Temafa)	1.35 (0.35)	1.61 (0.42)	2.41 (0.62)
5) Mechanically processed, fine fibers (Bahmer)	1.37 (0.36)	1.67 (0.43)	2.57 (0.67)
6) STEX**, fine fibers from 1) (Charle)	2.73 (0.71)	2.81 (0.73)	3.02 (0.78)
7) STEX**, coarse-medium fine fibers from 2) (Charle and La Roche)	3.15 (0.82)	3.22 (0.83)	3.43 (0.89)

* at an exchange rate of 1 DM = U.S. $0.57   nova 1996
** STEX = steam explosion process

they can be adequately refined by mechanical means; even then, the fineness and yield are lower as with flax.

Even when properly retted, field-retted hemp is usually too irregular to produce fine, uniform fiber bundles. In addition, the risk of bad weather and potential resulting crop damage is high; if weather conditions are unfavorable, it is even impossible to guarantee successful field retting.

Compared to flax, it is much more costly and risky to mechanically produce hemp fibers of equal fineness and uniformity. The price advantages of coarse-processed hemp fibers are lost after further refining, so flax can provide better quality at lower prices.

Finally, mechanical processing alone is not sufficient to allow the entry of hemp into higher-valued markets; with hemp, the required use of chemical and chemical-physical processes is much greater than with flax.

■ *High-Value Fine Fibers Made by Chemical-Physical Processing*

Fine, "mass cut" fibers usable for diverse purposes can be derived from coarse hemp-fiber bundles with the aid of various chemical or chemical-physical processes, so that industries can be provided with fibers that meet predetermined quality standards. These processes are largely independent of climatic conditions at the time the crop is cultivated. At the turn of the century, growers and manufacturers had already gained extensive experience in the "cottonization" of hemp tow. Currently, the decorticated whole fibers are used and, due to significantly improved processing techniques, show considerably better and more consistent fiber quality. Usage of the least problematic chemicals available, as well as recycling and treatment of the waste water, mean that new, environmentally sound techniques can be employed.

Of the three chemical-physical methods developed in Germany, one (the Flasin Process in Glachau) is in the industrial implementation phase, another (Ecco Ultrasound Processing in Bavaria) is in the pilot stage, and the third (Reutlingen Steam Explosion Processing, or STEX) has already been thoroughly industrially tested. STEX, developed at the Institute for Applied Research (IAF) at FH Reutlingen, is described in detail below.

■ *Steam Explosion Process Technology*

For this process, decorticated and well-cleaned fibers are used as raw or input materials. The measurable fiber-quality characteristics (degree of maturity and retting, chemical/morphological and technological characteristics) are determined to achieve optimal management of the processing parameters. Depending on the quality of the fibers and their intended usage, the processing parameters are adjusted, corresponding to a "product-property design."

The chosen starting material is pretreated with a waterproofing solution and exposed to intense steam in a reactor similar to a pressure vat. Depending on fiber quality and subsequent intended use, pressure (0–12 bar, 0–175 psi), time (1–30 minutes), and alkali concentration are adjusted. After the required reaction time, normal pressure is achieved by means of a release valve ("steam explosion"). Through this sudden pressure change, the developed fibers are shot into a catchment vessel via a pipe system. Further refinement of the fiber bundles occurs with the sudden reduction of pressure and subsequent acceleration of the fibers in the pipe system. Chemical substances that were decomposed during steam processing are flushed out during posttreatment when fibers are washed, bleached, or finished. Depending on the equipment of the plant, standard machines used in the textile industry can also be utilized here. After drying and, if necessary, a preopening process, the fibers can be further processed (within the spinning industry, nonwovens industry, etc.).

■ *The Economics of the Steam Explosion Process (STEX)*

The price for STEX-treated fibers is DM 2.70 to 3.40 per kilogram (U.S. $0.70–$0.90 per pound), which is definitely higher than the price for mechanically developed fibers. This is because the potential for high-quality uses is significantly improved.

Chemically-physically processed hemp is more economically advantageous than flax, because the decorticated, coarse fiber bundles can be obtained less expensively. The fineness of hemp fibers does not quite reach the quality of flax fibers, but the hemp fibers still command an attractive price. The prices for chemically processed hemp fibers can compete with those for cotton for open-end spinning mills, or with the prices of glass fibers for fiber-reinforced synthetics. These prices are relatively stable with regard to any fluctuation in the price of raw materials (hemp stalks). Even with a complete end to EU subsidies, the price change would be comparatively small: from DM 3.00 to DM 4.00 per kilogram (U.S. $0.80–$1.05 per pound), which could be offset by the optimization of processes and price leeways in the high-priced fiber markets. Typical areas of use for these high-priced fibers are:

- ■ apparel textiles (hemp-cotton blends from open-end (OE) spinning mills)

- ■ mats for thermal insulation, used in construction

- ■ fiber-reinforced synthetic materials (e.g., substitutes for fiberglass)

- ■ cellulose (with subsequent refinement)

The only economic problem is that the coarse fibers can be transported more cost-effectively over greater distances than can hemp stalks. Should the price of hemp stalks and consequently of coarse hemp fibers increase as the result of a reduction in EU subsidies, it would be financially worthwhile to import hemp tow from eastern Europe for chemical processing.

The short- to medium-term construction of chemical-physical fiber-processing facilities will play a key role in the German hemp economy, because only with those methods can fine, standardized fibers for high-value product lines be produced from mechanically processed hemp fibers.

■ The Structure of Fiber-Processing Plants

If it can correctly be assumed that the two processing methods discussed above will be primarily implemented, the structure of the fiber-processing plants will follow automatically.

The high transport costs for hemp stalks render regional development at the first processing stage a necessity. Each hemp-growing area encompassing 1,000–1,500 hectares (2,500–3,700 acres) will ideally be situated within 50 kilometers (30 miles) of a mechanical fiber-processing plant. The processing plants separate the hurds from the fiber, and are equipped with machinery that can produce pre-, coarse, or possibly even medium-grade fiber. At the same time, the hurds are collected, refined, and marketed regionally. Further manufacturing of the fibers to nonwovens—the initial stage for many product lines—in the processing plant is favorable.

The fiber-processing plants can sell coarse to medium-fine fibers directly to other industries, as well as to other fiber-processing plants that more extensively chemically-physically refine the fibers. These other facilities can either be located next to and supplied by the processing plants or can be in the immediate vicinity of the customer base. In the second instance, the facilities are supplied nationwide by several mechanical processing plants. The transport costs for coarse, decorticated fibers are minimal, and allow for transport over several hundred miles.

9.3 Favored Product Lines

Whether hemp will have a future in Germany as a renewable resource is primarily dependent upon how soon more extensive product lines can be created. These should require only minimal periods of time from cultivation, harvesting, and fiber processing to the production of marketable end products. These product lines must be technically feasible, have a sufficient value, offer environmental advantages, and find their niche in the market.

The Hemp Product Line Project (HPLP, nova 1996), conducted by the nova Institute in cooperation with the IAF/FH Reutlingen and the ifeu Institute Heidelberg, investigated which hemp product lines have the best chances to be realized in Germany from a technical, economic, and environmental standpoint. The eleven most highly favored product lines determined by this project are summarized below. In addition to technical, qualitative, economic, and environmental aspects, market structure and short- to medium-term market volumes were considered.

Besides these favored product lines, there are others that were of historical significance to hemp, but that no longer play a relevant role because of technological progress (e.g., hemp rope in the shipping industry). There are others that may be of great interest in the future but that are not currently favored because, according to scientists involved in the HPLP project, sufficient implementation seems possible only in the medium to long term.

Calculation of the market potential of the favored product lines described below resulted in a potential nationwide hemp-growing acreage of 30,000 hectares (75,000 acres) per year. Seed and oil product lines are essentially by-products of dual usage, thus they would require no additional acreage.

9.3.1 Fiber Product Lines

■ *Textiles for Apparel and Cottonized Hemp*

The HPLP analyses indicate that one of the more interesting prospective product lines for German hemp is in the apparel sector. This will surprise many who view the potential demand for bast fibers as being exclusively in the technical sector.

The recommendations do not aim for a revival of traditional long-fiber textiles, which are already established on the market and could capture and occupy only the smallest niche markets. (Furthermore, the fibers would need to be imported from eastern Europe and China.) Much more interesting for Germany is the potential production of "cottonized" hemp fibers, refined by the use of modern chemical-physical methods. The technical characteristics of these fibers are so similar to those of the dominating cotton fibers that the hemp fibers can be processed on highly productive cotton spinning machines (open-end or OE spinning mills). In hemp-cotton blends it is possible to produce fine yarns of a quality between Nm 20 and Nm 40. Such yarns and fabrics have been produced in Germany in small quantities since 1996, and their technical feasibility has been proven.

Economic analyses show that such hemp fibers can be produced in Germany at prices competitive to cotton. Even without any EU subsidies, the price of hemp fiber remains similar to that of cotton. An ecological life-cycle assessment, from cultivation to the manufacture of jeans, indicates significant benefits from hemp in comparison to cotton—provided that the textile-processing machines can be modified to achieve the same productivity level (and thus energy use per garment) with hemp as they do with cotton.

The realization of this innovative textile line is primarily an investment problem. The needed mechanically processed hemp fibers became available at the end of 1996; the necessary chemical-physical processing technologies are at a stage where industrial implementation may constitute only minimal technical, economic, and ecological risks (one processing technology is already in the process of industrial implementation); and, finally, a market for innovative hemp fibers (at the same price as cotton) appears to exist. What is still missing is an investor or investors who will make a chemical-physical hemp-fiber refining process a reality. Possible methods for this include the Flasin Process, the Reutlingen STEX Process, and the Ecco Ultrasound Process.

Market potential is estimated as follows: in the short to medium term, cottonized hemp fibers can replace roughly 5% of domestically processed cotton, mostly in the form of hemp-cotton blended fabrics. In addition, approximately the same amount can be projected for export and processing, e.g., in eastern Europe. This leads to an estimated land usage of about 13,000 hectares (32,000 acres). In the long term, significantly larger growing areas are possible if the product lines prove to be successful on the market.

However, the use of chemically-physically refined fibers is not limited to the textile industry, but also exists in the technical marketplace, such as with substitutes for glass and mineral fibers. The industrial use of bast fibers as a substitute for glass fibers in fiber-reinforced synthetic materials is still in its infancy and is, in spite of technical feasibility and environmental advantages (as well as certain economic benefits), not expected to achieve relevant industrial implementation in the short to medium term without further development. However, thermal insulation (see next paragraph) offers economic realization in the short term.

▪ Mats for Thermal Insulation in the Construction Industry

One sector showing good potential for the short term is that of mats for thermal insulation. In Germany, as a result of increasingly strict regulations concerning heat insulation, the market for insulation materials is continuing to expand. Within the market, the area of environmental insulation materials is experiencing phenomenal growth rates.

Although thermal insulation mats from mechanically processed hemp fibers, of the kind presently available on the market, cannot match the insulating quality of mineral-fiber mat, the possibility for such a match exists with chemically-physically refined hemp fibers (insulating values R13 and R19). Hemp insulating mats can substitute for mineral-fiber mats without the need for structural modifications to a building. Facilities in which hemp-fiber mats for insulation can be produced have already been developed for other natural fibers. Hence, the technical feasibility of this usage has already been demonstrated.

The substitution of glass and mineral fibers by hemp has significant environmental advantages when primary energy usage and the potential for global weather change are considered. In addition, there are advantages to subsequent waste management (composting), and also to worker health from installation of the mats (a lower risk potential for lung cancer).

From an economic point of view, such substitution is not as viable. As long as no environmental tax reform is introduced to economically aid hemp's revival, hemp insulating mats will remain more expensive by a factor of 4 compared to inorganic fiber mats. That is why the use of hemp insulation will continue to be restricted to the environmental building sector in the short to medium term. In this application, hemp can economically compete with other natural fibers such as wool, cotton, and flax.

Environmental insulation materials presently have a market share in Germany of approximately 3% of the entire market for insulation materials, and a market expansion of approximately 6% is expected in the short to medium term. Roughly 40% of this market will be covered by cellulose insulation materials, the most economical environmental insulation material, which will successfully defend its market share in the roofing insulation. It is assumed that chemically-physically processed hemp fiber can secure one-third of the remaining market, which would equate to 400,000 cubic meters (about 10,000 metric tons of fiber or 11,000 short tons) or a corresponding area of 8,000 hectares (20,000 acres). Both prices and the marketing structure of environmentally friendly construction materials are presently an obstacle to obtaining a significantly larger market share.

Although a multitude of other fiber product lines (such as fiber pressboards from mechanically processed hemp fibers) are feasible in the construction industry, these were not favored in the Hemp Product Line Project, because mechanically processed coarse hemp fibers are expensive compared to wood fibers and because the superior qualities of hemp fiber products are less beneficial to the construction industry.

All other favored product lines require no chemically-physically processed hemp fibers. Coarse to medium-fine fibers that are clearly lower in price (and quality) compared to chemically-physically processed fibers are sufficient. Fibers of higher or lower quality can only be used efficiently in the cellulose industry, provided there is a method of entry into the processing chain.

■ Specialty Pulp and Paper for Technical Applications

Domestic hemp used in the production of bulk pulp for printing, writing, and packing paper will be only minimally used in the foreseeable future. The greatest hindrance to its use is a lack of economic competitive advantage compared to wood fibers. Presently, conventional methods of cellulose production from hemp fibers are more expensive by a factor of 4–5 than producing high quality sulfate (strong) cellulose from conifers. In addition to the higher cost of raw materials, cellulose production facilities with a capacity of 200,000–500,000 metric tons (220,000–550,000 short tons) per year, which is currently customary in the bulk cellulose industry, would encounter massive logistical problems if hemp were to be used. This is primarily due to transport and storage of the voluminous stock that is harvested annually. Various innovative techniques meant to reduce the price difference even for smaller production units are currently being investigated, for example in the Netherlands. Even if in the medium term these new methods are successful in producing the more expensive (factor 2–3) hemp cellulose, in the current paper industry they will be able to gain a foothold only in the "enthusiast" markets. However, due to a high total market potential, this could in the long term translate to a considerable quantity of hemp.

Significantly better short-term possibilities exist for use in the higher-priced sectors such as specialty pulp and papers. It is here that hemp cellulose—due to its high tensile strength, wet tear resistance, porosity, durability, and opacity—can technically and economically compete with the dominant specialty cellulose from cotton, flax, and abaca. Currently, more than 95% of the hemp crop cultivated in western Europe is used as specialty paper.

Of the German use of 33,000 metric tons (36,000 short tons) per year of pulp from fiber plants, only a relatively small proportion is currently occupied by hemp, roughly 700 metric tons (770 short tons) per year. Increased use would primarily diminish the use of cotton-linters and abaca. Several well-known German manufacturers of specialized paper are currently investigating from a technical and economic perspective which usages can feasibly be substituted by hemp cellulose. Preliminary results suggest that hemp may have more tensile strength in comparison to cotton-linters. A greater dependability on supplies of raw materials from domestic sources is another argument cited by integrated facilities within the specialty pulp and paper industry. These plants mechanically and chemically process decorticated fibers in-house for the production of specialized paper products. The environmentally sound use of hemp in this case especially demands a multistep system to treat the organically contaminated wastewater.

Discussions with German specialty pulp and paper manufacturers suggest that the short-term market potential for domestic hemp will primarily depend upon the intensity with which the industry tests the technical feasibility and promotes the implementation process. Because of the extremely limited availability of data regarding present production quantities and markets, only rough estimates of the potential of hemp for the paper industry can be made. These estimates suggest production of about 3,000 metric tons (3,300 short tons) of hemp fiber cellulose per year, with a corresponding total acreage of 3,500 hectares (8,500 acres).

■ Press-Molded Interior Panels for the Automotive Industry

Fibers from annual plants—particularly flax, sisal, jute, and now hemp—are increasingly used in place of wood fibers in the manufacture of press-molded interior panels for automobiles. Typical uses include door interiors, dashboards, and headrests. Such usage is made even though wood fibers are significantly less expensive than fibers obtained from annual plants. The reasons can be found in the qualitative advantages of fibers from annual plants:

- weight reduction
- positive results in crash tests
- higher stability

▪ the possibility of manufacturing more complex components from one raw material and in one step

Hemp fibers that have been subjected to mechanical coarse processing have proven to be well suited for manufacturing press-molded panels. Hemp fiber has high tensile strength and low weight. In addition, due to its high yield per acre it can be produced at comparatively favorable prices of about DM 1.00 per kilogram (U.S. $0.25 per pound). Considering hemp's additional environmental advantages, nothing should stand in the way of a promising future for hemp in the automobile interior industry.

In this low-price market, the fact that production costs for the required hemp fibers will react fairly sensitively to EU subsidies for fiber plants may prove to be a problem. Hemp will remain competitive in comparision to flax, but not in comparison to jute and sisal imported from developing countries unless technical and economic advances in mechanical fiber processing are made. It seems that only when such environmental considerations as pesticide use and methods of transport have more importance in economical considerations will hemp cultivation for this product line be viable without EU subsidies.

Finally, a fundamental political decision needs to be made regarding whether plant fibers from EU countries should be used in the manufacture of automobiles or whether this industry—in spite of all the environmental and structural advantages of supply from EU fibers—should be supplied with imported fibers only. If a suitable framework for the use of EU fibers is created, hemp, due to its good quality and high yield per acre, has the strongest likelihood of all European fiber plants of being used in the manufacture of automobiles.

With regard to short- to medium-term markets, presently about 4,000 metric tons (4,400 short tons) of fiber from annual plants are used by the German producers of automobile interiors. In the coming years, an increase to nearly 10,000 metric tons (11,000 short tons) is expected. Should EU subsidies remain stable or be slightly reduced, it is assumed that hemp can cover 40% of the amounts cited above, which would occupy a planting area of roughly 2,500 hectare (6,200 acres) per year.

■ *Geotextiles for Erosion Control*

Huge quantities of textiles—particularly mats and felt—are used for erosion control. Until recently, these have been almost exclusively produced from synthetic fibers. Lately there has been a growing interest in the use of biodegradable geotextiles from natural fibers in certain areas. Typical prospective uses are the protection of embankments and other such areas from erosion, the restoration of ski areas and slopes, and even the use of natural fibers as a greenery mat or felt embedded with plant seeds.

Experiments conducted up to this point—mostly with flax or jute—to replace synthetic fiber mats with natural fiber mats have been quite positive. The tested mats and felts deliver characteristics comparable to those of synthetic fibers. As a result of more stringent environmental regulations and a greater concern for the environment, it will become increasingly appealing to use biologically degradable mats and felts, especially when the geotextile must exert its function only over a specified amount of time—e.g., until roots hold the soil together.

Nonetheless, since fewer than 1% of the textiles for erosion control are now manufactured from natural fibers, there is at present an insufficient understanding of their use among civil engineers. New teaching materials for civil and hydraulic engineering refer to the use of natural plant fibers such as flax, jute, and coconut palms—an indication that people are actually beginning to change their views toward natural plant fibers and their potential uses.

Because of its high tensile strength (even when it's wet), minimal stretching, excellent water-absorption qualities, and comparatively slow biodegradability, hemp is an outstanding choice for the production of geotextiles. Compared to flax, which has a similar quality profile, the lower price of mechanically processed coarse hemp fibers is an important advantage.

When assessing the market potential for hemp, analysts have been very conservative in their assumptions. In at least 5% of all geotextile applications, synthetic fibers could be substituted with natural fibers. It is assumed that, under suitable conditions such as appropriate environmental support and education of civil engineers, in the near future 5% of the market for natural fibers can be secured, with hemp fibers accounting for 25% of the natural fibers market. This estimate requires 4,000 metric tons (4,400 short tons) of hemp per year and a corresponding acreage of 3,000 hectares (7,400 acres). This overly conservative estimate nonetheless shows the great potential for hemp fibers in the long term.

■ Needle-Punched Carpeting

Environmentally friendly carpets made from natural plant fibers play a subordinate role in the carpeting market, being predominantly purchased by private buyers in the so-called "eco-"sector. Woven and spun carpets are presently the only available variety of these eco-carpets; needle-punched carpets from natural fibers are not yet on the market.

Recent tests in the carpeting industry show that it will be possible to produce high-quality needle-punched carpets using mechanically processed hemp fibers. Such carpets stitched from hemp fibers could —due to their significantly lower price—create entirely new markets for carpets manufactured from natural fibers. This would also result in a total increase in market share.

Aside from the private sector, hemp needle-punched carpets are also suited for industrial use, e.g., in trade fairs and exhibits. Low-quality carpets and tapestries are used almost exclusively in trade fairs and exhibits; some of these must be disposed of after a single use. Since most contain polyamide or polypropylene, mass disposal of such carpets and tapestries is problematic and represents a significant cost factor.

A cost-competitive carpet made from hemp, which fulfills the wear-and-tear stability requirements and can be composted after use, may lead to increased interest on the part of both residential and commercial consumers.

Adequate marketing of cost-competitive hemp needle-punched carpeting and its fulfillment of the wear-and-tear requirements can boost the natural carpet market and increase its annual volume by 50%, or to about 10,500 metric tons (11,600 short tons) per year. Of this, about 3,500 metric tons (3,900 short tons) or 2,500 hectares (6,200 acres) could be secured by hemp. This amount corresponds to about three times the annual volume of needle-punched carpeting used by a large trade-fair construction company.

9.3.2 Hurd Product Lines

■ *Hurds Used as Animal Bedding*
Hemp hurds are obtained during every stage of mechanical process-ing, and must be sold for the highest price possible in order to render the processing plant profitable. The best short- to medium-term mar-ket appears to be in the area of animal bedding, particularly for horses.

Hemp hurds are especially well-suited for this usage because of their high absorption potential and easy compostability. Other favor-able characteristics include their light color and ease of distribution, as well as their suitability for sensitive horses that cannot tolerate straw due to allergies or respiratory diseases. Animal bedding made from hemp hurds proves to be as suitable as wheat straw, and provides similar in-sulation and softness. Animals easily adapt to such bedding.

From a cultivable area equalling about 30,000 hectares (74,000 acres), roughly 90,000 metric tons (99,000 short tons) of hurds accu-mulate during fiber processing. These are then separated into three qualitatively different types: horse bedding (36,000 metric tons, or 40,000 short tons), bedding for smaller animals (27,000 metric tons, or 30,000 short tons), and cat litter in pellet form (13,500 metric tons, or 15,000 short tons). The lighter-colored small-animal bedding can also be marketed as horse bedding.

The hurds for horse bedding are very easily marketed. Hemp hurds cannot be sold for as great a profit in rural areas, where straw com-petes, but they can readily compete with alternative bedding sources such as wood shavings (which have a market share of 10-40%) and peat, straw pellets, sand, etc. (about 10%). These alternative sources are pri-marily used for animals sensitive to straw, kept near cities, or kept in areas where straw is rarely available and costly to dispose of. In such areas, hurds will have a good market potential.

In France, England, the Netherlands, and Germany, hemp hurds have already been successfully introduced as animal bedding. Hurds from France and the Netherlands are being marketed in Germany, and the marketing of hurds from German farms is anticipated in 1997.

The German animal-bedding market is so large that it should easily take up the hurds from 30,000 hectares (33,000 acres). If hurds of all three quality types that are suited for horse bedding are sold in the horse-bedding sector, this would equal roughly a 15% share of bedding alternatives, or a mere 6% share of the entire horse-bedding market. Therefore it is useful and important for the operators of a fiber-processing facility to make sales deals with riding stables and horse boarders in regional markets.

The markets for small-animal bedding and cat litter are considerably smaller than the horse bedding market, and sales to these two sectors are much more difficult to make. In any case, a nationwide marketing campaign seems necessary. With a solid sales structure, such a campaign will successfully compete against a variety of products currently on the market. One marketing potential of which full advantage has not yet been taken is eco-stores or health food stores, which presently have no competing products available. Good product quality, steady supply, decent compostability, and a competitive price should ease the entry of hemp hurds into this market.

Other potential purchasers are animal owners and zoos. Since hurds, due to their minimal density, have high transport costs, their use is primarily dependent upon local demand.

The use of hemp hurds as insulating and building materials in the construction industry (often described as having a promising future) was not favored in the Hemp Product Line Project. The reason: aside from special applications, hemp hurds, in comparison to other resources for the construction industry, display no especially outstanding qualities. This holds true for hemp hurds as insulation, in which usage they are inferior to the more customary environmentally friendly insulating materials made from cotton or waste paper.

Since competing resources are sold at a wholesale price of DM 100 per metric ton (U.S. $50 per short ton), hurds will obtain similar prices on the market. This price, however, would not be lucrative, because in the horse-bedding market the hurds can obtain a price of DM 200–300 per metric ton (U.S. $100–160 per short ton) after being processed.

9.3.3 Seed and Oil Product Lines

The success of product lines based on hemp seeds and hemp oil are the greatest surprise since the rediscovery of hemp as an industrial crop. Numerous businesses have already introduced foods and body-care products, furniture oils, and dyes, as well as soaps and detergents, to the consumer market. Aside from the high value of the newly discovered resources and the strong interest on the part of consumers, this success is primarily due to the fact that hemp seeds and hemp oil can be manufactured in existing factories and integrated into existing production lines without major problems. In the HPLP study, the favored potential uses included foods, bodycare products, and suppliers of the essential nutrient gamma linolenic acid (GLA), in which hemp oil is rich.

In Germany in the short to-medium term, hemp seeds will be cultivated along with fiber hemp as a dual-usage crop. The yield per acre for hemp seeds is below that of most commercial oil-producing plants. The main reason for this is that there are currently no fiber-hemp varieties optimized for high seed yield. In addition, there are as of yet no early maturing varieties that would produce an abundance of mature seeds in the western European climate. Such plants would have similar yields to flax and, because of their high-quality oil, could conceivably achieve equal importance.

▪ **The Food Sector**

Hemp seeds are a high-quality nutritious food. As nourishment, they contain a wide spectrum of fatty acids and amino acids. The seeds contain large amounts of monounsaturated or polyunsaturated fatty acids, mainly linoleic acid (60%) and linolenic acid (20%). The proportion of polyunsaturated fatty acids to saturated fatty acids is nutritionally and physiologically favorable. Hemp oil's content of valuable gamma linolenic acid (GLA) ranges from 2–4%+, and hemp seeds contain all of the essential amino acids in an easily digestible form.

The high quality of hemp oil is equal to that of hemp seeds, and this oil is already available on the market in the form of salad oil. After pressing of the seeds, a very nutritious low-fat seed cake remains that can be used for a variety of purposes, including hemp flour, beer, and animal food supplements.

Hemp seeds have been a traditional source of nourishment in many countries. The seeds can be roasted, ground, or used as hemp flour in a variety of ways, such as in breads, crackers, cakes, and pastries. Other possible uses include hemp milk and tofu and hemp spreads for bread and crackers, and the sprouted seeds can be added to salads.

The HPLP study (nova 1996) specifically investigated the potential role of hemp seeds in the German grocery sector in the short-to-medium term. Given the current competitive situation on the international market, high-quality hemp seeds from controlled biological cultivation stand the best chance of success. Such biologically cultivated seeds are expected to attain a wholesale market price of about DM 2.00 per kilogram (U.S. $0.50 per pound), so the seeds will effectively compete with comparable imported seeds. For most grocery items in which hemp seeds can be possible ingredients, the actual price for the raw material plays a minor role.

The situation appears to be different for hemp oil, where the seed price directly determines the oil price. The European wholesale price for cold-pressed hemp oil is DM 20–25 per liter (U.S. $45–$55 per gallon), and this price may remain that high as long as seed prices do not decrease due to better hemp varieties. In the United States, the price has fallen to $25–$30 per gallon. Because of the polyunsaturated fatty acids in hemp oil, available in natural oil as triglyceride, hemp oil is easily digested. The market for hemp oil in food usages is currently limited to the high-quality sector of salad oils with certified organic quality. Additional markets can be gained in the area of therapeutic oils (see below).

It can be assumed that, in Germany, hemp seeds will continue to be cultivated as a dual-usage crop with fiber hemp. If dual usage is successfully carried out on one-third of the acreage (10,000 hectares, or 25,000 acres), roughly 6,000 metric tons (6,600 short tons) of seeds (equal to 1,500 liters or 400 gallons of oil) will be marketable as food products or supplies to either of the two groups discussed below (producers of natural bodycare products and GLA suppliers). The most important factors in the food sector are the high quality of the seeds and their harmlessness with regard to intoxicating effects, which latter factor (because the seeds contain no cannabinoids) can be advertised to producers and consumers. The increased media interest in hemp products can be very useful in this respect.

The acreage used solely for seed production is primarily dependent on the cultivation of early-maturing and high-seed-yielding varieties of hemp, as well as on the further progress of harvesting techniques that determine seed yield, profitability, and seed prices.

■ *Natural Bodycare Products*

Because of its high quality, hemp oil is well suited for use in natural bodycare products in which only oils with a high food-grade quality are used. It is also well suited for therapeutic use, and is particularly a component in natural products used in the care of sensitive or troubled skin. The reason for this is the excellent spectrum of fatty acids, including gamma linolenic acid (see below).

Regionally, seed hemp can be grown and the seeds processed according to cooperative agreements. In this way, the producers can guarantee to purchasers that no undesirable substances are used during any of the processing. Fortunately, hemp can be grown for certified organic quality at relatively low expense. A high-quality, regional, certified organic hemp oil will certainly find its place in the local natural bodycare markets.

Quantitatively, this market potential is estimated to be less than that of the food sector, because the natural bodycare sector requires comparatively little oil for its products.

■ *Gamma Linolenic Acid in the Cosmetics and Pharmaceutical Industries*

Hemp oil is one of the few edible oils that contain the rare gamma linolenic acid (GLA), which is used in both the pharmacological and the cosmetics industries. The most important use is in the area of chronic skin disorders such as neurodermatitis. External (with ointments) or internal (with capsules) treatment with gamma linolenic acid can balance a lack of essential fatty acids and return the moisture loss of the skin back to normal hydration. Other uses for GLA include treatment of rheumatoid arthritis, diabetic neuropathy, and premenstrual syndrome.

Current GLA sources are evening primrose and borage. Both are specialty crops, and their cultivation and harvesting is labor-intensive. The use of agricultural chemicals is customary, and harvesting is generally done by hand. These plants are not cultivated in Germany; the oil is imported primarily from New Zealand.

Even hemp varieties presently available that do not have high seed yields produce a hemp oil that, calculated at the same GLA content, is less expensive than evening primrose or borage oils. In addition, hemp oil can be integrated into normal meals, while evening primrose and borage oils are not particularly tasty and have therefore no application as food oils.

In order to introduce hemp oil into medicinal applications, it would be necessary to increase the GLA content from the present 2–4% level to about 10%. Hemp oil with a 10% GLA content could immediately replace evening primrose and borage oils.

Only rough estimates of the market potential for GLA can be made, because the GLA market cannot be statistically quantified. In Germany, at least 40 metric tons (44 short tons) of evening primrose and borage oils are used each year in the bodycare and pharmaceutical industries. If a low-cost method of increasing the GLA content in hemp oil were realized, hemp oil could gain this entire market. It is estimated that approximately 150 metric tons (170 short tons) per year of hemp oil would be necessary at this higher GLA content, which would require an acreage of 500 (seed cultivation) to 1,000 (dual usage) hectares (1,250–2,500 acres).

■ *Natural THC-Based Therapeutic Drugs*

In Germany as elsewhere, administrative measures currently impede the use of medication with natural THC-based pharmaceuticals. Natural cannabis products containing THC are not legal to transport or prescribe according to current German drug laws (BtMG). Therefore there is no legal market for such drugs. However, in 1996 the responsible committee of experts considered reclassification of the qualitatively and quantitatively most important pharmacological ingredient of the cannabis plant, Delta-9-THC, in the BtMG so that THC could soon be legally prescribed. According to the latest reports from the Department of Health, adoption of such an ordinance by the federal government should take effect in the first half of 1998. However, the position of THC in the framework of the federal government will be expressly amended with the prefix "synthetically produced," in order to reject the manufacture of THC preparations from the buds of THC-rich cannabis plants with an identical molecular structure of (–)trans-Delta-9-THC. This is in spite of the fact that such natural prescriptions are less costly to produce and are requested by patients.

Presumably, a report from the Federal Institute for Pharmaceutical and Medicinal Products dating from the fall of 1995, which was published in June of 1996 in the National Health Brief entitled "The Medicinal Use of Cannabis Products," contributed to the decision to recommend allowance of the medicinal use of THC. As a result, an uncritical euphoria with regard to the therapeutic possibilities of cannabis or THC followed. On the other hand, ungrounded opposing positions expressed a general disapproval, claiming that "there are better therapeutic alternatives for every application." Potential therapeutic uses for THC include treatment for lack of appetite; emaciation from AIDS and cancer; the nausea associated with chemotherapy; physically limiting spastic paralysis with paraplegia and multiple sclerosis; glaucoma; seizures; and chronic pain.

While the use of cannabis products for medicinal purposes (particularly in English-speaking countries) was very common in the nineteenth century, these products have been legally treated as hard drugs and publicly perceived as such during the past few decades. However, people are becoming increasingly aware that THC poses less potential for addiction than many legally prescribed sleeping pills and sedatives and can at the same time broaden the therapeutic spectrum of available medications.

The latest research also verifies the effectiveness of THC or cannabis products containing THC at various stages of illness in patients who have previously not responded well to available therapies. A research project with hundreds of patients to investigate results of the use of natural cannabis extracts is scheduled to begin in several German, Dutch, Swiss, and Austrian clinics in 1998. Its goal is to license natural cannabis products for medicinal purposes.

Only a rough estimate can be provided of the status that THC will gain in modern medicine. Attempts at estimating the potential market share remain speculative for the most part. Small, well-protected plots of varieties with high THC content will presumably suffice to guarantee the necessary amounts of THC for pharmaceuticals. The potential estimated market volume in Germany for medications containing THC is more than DM 100 million (U.S. $57 million) per year.

9.4 Regulations in Germany and the European Union (1996 Status)

■ *A Discussion of Regulations in Germany*

Farmers who intend to grow low-THC fiber hemp must file a report to that effect at the Federal Institute for Agriculture and Nutrition (BLE) in Frankfurt, on the form entitled "Notice of the Cultivation of Hemp." On the same form, EU subsidies can be requested. Contact: BLE, Referat 321, Adickesallee 40, 60322 Frankfurt a. Main, Germany. Tel: +49-69-1564-386.

Anyone wanting to grow low-THC hemp, but not on a farm, or THC-rich cannabis, e.g., for research purposes, must consult with the National Opium Division of the Federal Institute for Drugs and Pharmaceuticals (Genthinerstrasse 38, 10785 Berlin, Germany). Approval will be granted only if the purpose of the proposed cultivation serves the public interest from a scientific perspective or otherwise.

The BLE distributes a leaflet on the cultivation of hemp and granting of hemp subsidies that contains excellent information. The leaflet is free of charge and can be obtained from the agency. Below are cited some of the most important statements from the leaflet (April 1996):

The cultivation of hemp is permitted only for agricultural businesses in accordance with Article 1, Section 4, of the law regarding the provision for the pension of farmers (ALG), whose acreage meets or exceeds the minimum size described in Article 1, Section 2 ALG. . . . Apart from the special regulation for beet farmers, agricultural businesses can cultivate hemp within the current EU regulations as:

- *A subsidy crop in the framework of the EU-subsidy regulations for hemp (1995 and presumably also 1996: DM 1,510.45 per hectare, U.S. $349.30 per acre)*

- *A renewable resource grown on land otherwise idle and receiving a subsidy (average DM 750 per hectare, U.S. $170 per acre);*

> *dual support from hemp subsidies and compensation for non-production is excluded*

■ *An agricultural crop with no subsidy regulation*

Any cultivation of industrial hemp is to be reported to the Federal Institute for Agriculture and Nutrition along with the enclosed form "Notice of the Cultivation of Hemp." . . . *For the cultivation of hemp, only certified seeds of the following EU-registered varieties may be used.* The leaflet then lists a total of fourteen registered French, Italian, and Spanish varieties of hemp. In 1996, Epsilon 68 and Santhica 23 were included as new varieties. *As proof of the use of certified seeds,* the leaflet goes on, *the official labels of varieties used have to be submitted with notice of cultivation.*

The application for EU-hemp subsidies is acknowledged only if the farmer:

■ *has submitted a declaration of cultivation to the Federal Institute for Agriculture and Nutrition*

■ *has submitted a subsidy application. . . .*

The subsidy will be granted for plots that were sown and harvested entirely. . . . The plots qualify as harvested only if harvesting follows seed formation, plant growth is complete, and the stalk, possibly even without the seed capsule, is used.

Seed formation is effectively complete if, in a representative test sample from the plot, it is conclusively shown that matured hemp seeds predominate. . . .

The mower should be adjusted to a maximum mowing height of 20 centimeters (8 inches). The Federal Institute for Agriculture and Nutrition is to be notified immediately before harvesting, in writing or by fax (Notification of Harvesting). In practice, the harvest should be reported fourteen days in advance. The stubble remaining in the field after mowing can legally not be plowed under until at least twenty days following the harvest. This period is necessary to allow for inspections.

The Federal Institute for Agriculture and Nutrition inspects the entire national cultivation of industrial hemp by means of spot checks. This also includes cultivation by farmers who have not applied for hemp subsidies. . . . Anyone not reporting the cultivation of hemp, or incorrectly reporting it, *whose report is incomplete, or who does not report in a timely manner is infringing the rules and can be fined up to DM 50,000 (U.S. $28,600) according to the German Drug Law.*

Finally, every producer of hemp is required to report the estimated average yield of straw, fiber, and seeds to the Federal Institute for Agriculture and Nutrition no later than October 15.

Important: The rule that harvesting may occur only when more than 50% of the hemp seeds have matured only refers to the granting of hemp EU-subsidies, not to the subsidy for officially idle land. Officially idle land on which hemp is cultivated is not liable to any restrictions during the harvesting period, but is liable for proof of usage. The permitted varieties of hemp comply with those on the EU list; their farming is not legally bound by an ordinance, but rather by the German Drug Law (BtMG).

▨ *A Discussion of EU Regulations*

Aside from a justifiable critique of the regulatory expenditures that the potential hemp grower has to bear, the following aspects of the EU regulations are particularly problematic:

At the beginning of 1996, representatives from Spain and France successfully argued for a change in the EU-subsidy rules [Regulation (EG) Nr. 466/96 of the commission dated March 14, 1996 for a change in Regulation (EWG) Nr. 1164/89: Brief dated March 15, 1996, Number L65/6], which provides that the subsidy will be granted only when harvesting is carried out after the seeds have matured. This is the case when the "definitively matured hemp seeds numerically prevail." It is not clear whether this new ordinance is simply a clarification of previous regulations or represents more stringent regulations. What led to this alteration were the concerns of Spanish and French hemp growers who believe that, as hemp cultivation in the EU increases, subsidies may be decreased in the near future.

For years it has been common knowledge that in Spain and England several 10,000 hectares (75,000 or more acres) of flax are cultivated and obtain EU subsidies, although the fibers are of no use, but rather are burned or plowed under. This practice endangers the current high flax subsidies, without which the countries (France, Belgium) that use the flax would particularly suffer. As far as hemp is concerned, there has been no subsidy misuse so far by "benefit bandits." However, the recent rapid expansion of growing plots in the EU (refer to Figure 1A) creates the fear that the hemp subsidies could also be reconsidered. To counter these legitimate concerns with a new regulation imposing an obstruction to hemp cultivation in central and northern Europe is no solution. It would be more sensible to make the EU subsidies dependent upon guaranteed purchase and processing. In June of 1996, the first discussions regarding such a strategy took place in Brussels, Belgium, but a revised regulation has not yet been passed.

In any case, the new regulation impedes the subsidizing of hemp cultivation in central and northern Europe, as well as the use of dioecious varieties of hemp. It is a well-known fact that presently EU-registered varieties of hemp in central and Northern Europe often only reach maturity in October; in cold, rainy years, only small portions may reach maturity. In many regions it is sensible to harvest the plants before the seeds mature, at the end of August or beginning of September, and to supply users at that time. According to the new regulation, this is possible only with the forfeiture of EU subsidies. Farmers are forced to continue growing the hemp plants through the end of September or even until the beginning of October, thereby taking significant risks regarding weather conditions.

Determining when to harvest "in accordance with EU-subsidy regulations" is based on the consolidation of initially separated subsidies for the cultivation of fiber hemp and seeds (as is the case with fiber flax and flax oil). Since both uses qualify for the subsidy payments, the successful production of straw and seeds was to be ensured with the establishment of a harvesting time. Another point in favor of establishing a harvesting time is to prevent the crop from being plowed under after a short period and not being used. In such cases no EU subsidies are granted (see above).

However, hemp cultivation in southern Europe will be unevenly favored by the new regulation. The motion made by the Dutch firm HempFlax, and backed by Germany and Finland, to also grant EU subsidies when the hemp plants are harvested before seed maturation, was rejected by the EU Commission "Advisory Committee for Flax and

Hemp," which is mainly influenced by individuals representing the interests of Spanish and French hemp agriculture. In October of 1996, representatives of hemp interests in Germany and the Netherlands met at the Ministry of Agriculture in Bonn, Germany, to work out a new proposal for EU-subsidy regulations. The new proposal aims at a discontinuation of the fixed harvest period, as well as a discontinuation of the need for proof of usage.

The French hemp growers who previously had a monopoly on EU-subsidized (monoecious) varieties of hemp will probably not be able to hold this enviable position much longer. Central and northern European countries will, as a result of the EU regulation, endeavor to register certain northern, early-maturing hemp varieties from such countries as Poland and Russia, which mature under less than optimal climatic conditions.

In addition, the new regulation impedes the use of dioecious varieties of hemp. Such varieties were commonly grown in Hungary and are among the highest-yielding varieties in the world. With dioecious varieties that have male and female plants, it is difficult to harvest after the seeds have matured. It is much more common to begin harvesting as soon as the male plants flower, which is generally in mid-August. Shortly thereafter, the male plants begin dying off, while the seeds from the female plants continue to mature over the next four weeks. The result of the new guidelines is that the required harvesting deadline leads to quality loss in the fibers of the male plants, due to lignification and over-retting.

Thus, it will be very difficult to implement an EU-subsidized harvest with dioecious varieties of hemp that results in high quality and high yields. This is a disadvantage to hemp cultivation in EU countries, because the dioecious Hungarian varieties such as Kompolti are the highest-yielding hemp varieties in the world.

Hemp Resources

Associations

Australian Hemp Industries
  Association
P.O. Box 236
New Lambton, New Castle
Australia 2305
Tel: 61-49-55-6666
Fax: 61-49-55-6655
Austhemp@hunterlink.net.au

Hemp Industries Association
P.O. Box 1080
Occidental, CA 95465
Tel: (707) 874-3648
Fax: (707) 874-1104
info@thehia.org
www.thehia.org

International Hemp Association
Postbus 75007
1070 AA Amsterdam, Netherlands
Tel/Fax: 31-20-6188758
iha@euronet.nl

Kentucky Hemp Growers
  Cooperative Association, Inc.
P.O. Box 8395
Lexington, KY 40533
Tel: (606) 252-8954

New Zealand Hemp Industries
  Associations, Inc.
P.O. Box 38-392
Howick, Auckland, New Zealand
Tel: 025-514-22
Fax: 04-477-4819
nzhemp@ex.co.nz

Associations – *continued*

North American Industrial
  Hemp Council, Inc.
P.O. Box 259329
Madison, WI 53725-9329
Tel: (608) 224-5135
Fax: (608) 224-5110
sholtea@wheel.datcp.state.wi.us
www.naihc.org

Government Licensing

Bureau of Drug Surveillance
122 Bank St., 3rd Fl.
Ottawa, Ontario K1A 1B9 Canada
Tel: (613) 954-6524
Fax: (613) 952-7738

Home Office
SE Region Drugs Inspectorate
50 Queen Anne's Gate
London SW1H 9AT England
Tel: 0171-273-3856
Fax: 0171-273-2671

United States DEA
Room W12058
700 Army Navy Drive
Arlington, VA 22202
Tel: (202) 307-7927
Fax: (202) 307-4502

Harvesting Equipment

Hempline Inc. *(see Processors
  listing)*

John Deere International
  (see your local dealer)

Organizations

GATE
Dr. Iván Bósca
3356 Kompolt, Hungary
Fax: 36-36-489000

HEMPTECH
P.O. Box 1716
Sebastopol, CA 95473
Tel: (707) 823-2800
Fax: (707) 823-2424
www.hemptech.com

nova Institute
Thielstr. 33
50354 Hurth, Germany
Tel: 49-2233-978374
Fax: 49-2233-978369
nova-H@T-online.de

Processors

Consolidated Growers &
   Processors Inc. (CGP)
P.O. Box 2228
Monterey, CA 93942-2228
Tel: (888) 333-8CGP
Fax: (888) 999-8CGP
www.congrowpro.com

Hemcore Ltd.
Station Road, Felsted
Great Dunmow, Essex
England CM6 3HL
Tel: 441-371-820066
Fax: 441-371-820069

HempFlax B.V.
Hendrik Westerstraat 20-22
9665 AL Oude Pekela
Netherlands
Tel: 31-597-615-516
Fax: 31-597-615-951
www.hempflax.com

Processors – *continued*

Hempline Inc.
632#1 Elizabeth St., London
Ontario N45W 3S7 Canada
Tel: (519) 434-3684
Fax: (519) 434-6663
www.hempline.com

Kenex Ltd.
R.R. #8, Chatham
Ontario, Canada N7M 5J8
Tel: (519) 352-2968
Fax: (519) 352-6667
www.kenex.org

Publications

See list at www.hemptech.com

Seed Cultivars

Ag Innovations
202 S. Westwood Blvd., Ste. 32
Popular Bluff, MO 63901
Tel: (573) 785-8711
Fax: (573)785-3059
www.pbmo.net/oxhemp

CGP *(see Processors listing)*

FIBRO-SEED Ltd.
GATE Agricultural Research
   Institute
Dr. Iván Bósca
3356 Kompolt (Heves)
H-3356 Hungary
Tel: 36-36-489-082
Fax: 36-36-489-000
gatefrki@gateki.ektf.hu

Kenex Ltd. *(see Processors
   listing)*

International Hemp Association
   (see Associations listing)

Bibliography

Abel, E. L. (1980): *Marihuana: The First Twelve Thousand Years.* Plenum Press, New York.

Berenji, J. (1992): Konoplja. Bilten za hmelj, sirak i lekovito bilje, 79-85.

Bócsa, I. (1961): Data on the seed yielding capacity of the unisexual F1-hybrid hemp. Növénytermelés, 10: 43-50.

Bócsa, I. (1966): Neue Richtungen und Möglichkeiten in der Züchtung des südlichen Hanfes in Ungarn. Acta Agric. Scandinavia, Suppl. 16, 292-294.

Bócsa, I., Pozsar, B., Majkó, Z. (1969): Die Züchtung einer hellstengeligen, südlichen Hanfsorte. Z. f. Pflanzenz., 62: 231-240.

Bócsa, I., Manninger, G. (1981): A kender és rostlen termesztése. Mezögazd, Kiadó, Budapest, 199. p.

Bócsa, I. (1995): Die Hanfzüchtung in Ungarn: Zielsetzungen, Methoden und Ergebnisse. Bioresource Hemp Symposium, March 2-5, 1995, Frankfurt/M., nova Institute/HEMPTECH.

Bredemann, G. (1924): Auslese faserreicher Männchen zur Befruchtung durch Faserbestimmung an der lebenden Pflanze vor der Blüte. Angew. Bot., 6: 348-360.

Bredemann, G. (1953): Verdreifachung des Fasergehaltes bei Hanf (*Cannabis sativa* L.) durch fortgesetzte Männchen- und Weibchen-Auslese. Mat. Veg., Den Haag, 1: 167-182.

Breitfeld, R. (1995): Rohstoff Hanf. Hanfgesellschaft, Berlin, 28. p.

Ceapoiu, N. (1958): Cinepa. Studiu monografic. Edit. Acad. R. P. R., Bucarest, 734. p.

Davidjan, G.G. (1971): Wirkung des Lichtes und Gibberellins auf Wachstum, Entwicklung und Geschlecht des Hanfes. Presentations at the International Hemp Research Conference, Kompolt, Hungary, July 28-31, 1970. Rostnövények, 41-59.

de Meijer, E. P. M., van der Kamp, H. I., Eeuwijk van, F. A. (1992): Characterisation of *Cannabis* accessions with regard to cannabinoid content in relation to other plant characters. Euphytica, 62: 187-200.

de Meijer, E. P. M. (1993): Evaluation and verification of resistance to *Melodogyne hapla* Chitwood in a *Cannabis* germplasm collection. Euphytica, 71: 49-56.

de Meijer, E. P. M. (1995): Fibre hemp cultivars: A survey of origin, ancestry, availability and brief agronomics characteristics. *The Journal of the International Hemp Association,* 2: 66-72.

Fédération Nationale des Producteurs de Chanvre (1995): La culture du chanvre: Battage en atelier. Le Mans.

Fédération Nationale des Producteurs de Chanvre 1995 La culture du chanvre: Battage sur champ. Le Mans.

Gáucá, C. (1995): Personal communication.

Gutberlet, V., Karus, M. (1995): Parasitäre Krankheiten und Schädlinge an Hanf (*Cannabis sativa* L.). nova Institute, Cologne, 57. p.

Hennink, S. (1995): Personal communication.

Herer, J., Brockers, M., Katalyse-Institute (1993): Die Wiederentdeckung der Nutzpflanze Hanf. Zweitausendeins, Frankfurt/M.

Heuser, O. (1927): Die Hanfpflanze. In: Herzog, O. (Ed.): Hanf und Hartfasern. Verlag Julius Springer, Berlin.

Hingst, W., Mackwitz, H. (1996): Reiz-Wäsche. Campus-Verlag, Frankfurt, 208 p.

Hoffmann, W. (1946): Helle Stengel - eine wertvolle Mutation des Hanfes (*Cannabis sativa* L.). Der Züchter 17/18: 56-59.

Hoffmann, W. (1957): Flachs- und Hanfbau. Deutscher Bauernverlag, Berlin.

Hoffmann, W. (1960): Hanf: *Cannabis sativa* L. In: Kappert-Rudorf: Handbuch der Pflanzenschutzzüchtung, 2nd Ed., V: 204-264.

Höppner, F., Menge-Hartmann, U. (1995): Cultivation experiments with two fibre hemp varieties. The Journal of the International Hemp Association, 2: 18-22.

Karus, M. (1995): Hanf - Biorohstoff mit Zukunft. nova Institute, Hürth.

Karus, M. (1996): Hanf in Deutschland 1996 - das erste Jahr. 4th Ed., nova Institute, Hürth.

Köber-Grohne, U. (1988): Nutzpflanzen in Deutschland - Kulturge-schichte und Biologie. Konrad Theiss Verlag, Stuttgart.

Kozlowski, R. (1995): Personal communication.

KTBL (1994): Taschenbuch Landwirtschaft, Darmstadt.

Kuntze-Rechman (1988): Schwertfeger Bodenkunde, Verlag Eugen Ulmer, Stuttgart.

Lohmeyer, D. (1995): Personal communication.

Lohmeyer, D. (1996): Die Hanfernte - Statusbericht und Ausblick. nova Institute, Hürth, 24. p.

Martinov, M., Berenji, J., Herak, S. (1994): Spremanje konoplje za papirno vlakno. Revija poljopr. techn., 4: 24-27.

Mathieu, I. P. (1995): Personal communication.

Neuer, H., Sengbusch von, R. (1943): Die Geschlechtsvererbung bei Hanf und die Züchtung eines monözischen Hanfes. Züchter, 15: 49-62.

nova (1995): *Bioresource Hemp: Proceedings of the Symposium.* Bio-resource Hemp Symposium, March 2-5, 1995, Frankfurt/M., nova Institute/HEMPTECH.

nova (1996): Das Hanfproduktlinienprojekt. Supported by Deutsche Bundesstiftung Umwelt, Badische Naturfaseraufbereitung, hanfnet Hannover, and TreuHanf Berlin. Conducted by nova Institute Hürth, IAF/FH Reutlingen and ifeu-Institut Heidelberg.

nova (1997): *Bioresource Hemp: Proceedings of the Symposium.* 2nd Bioresource Hemp Symposium, February 2 - March 3, 1997, Frankfurt/M., nova Institute/HEMPTECH, 2nd Ed.

Roulac, J. (1997): *Hemp Horizons: The Comeback of the World's Most Promising Plant.* Chelsea Green Publishing Company, White River Junction, Vermont, 212 p.

Themagroup Regionale Ontwikkeling (1982): Andere gewassen in de veenkoloniale akkerbouw - mogelijkheden voor bouwplanver-ruiming? Wageningen, The Netherlands.

van der Werf, H. (1994): Crop physiology of fibre hemp (*Cannabis sativa* L.). Doctoral thesis, Wageningen.

von Buttlar, H. B. (1995): Personal communication.

Wirowetz, V. (1995): Personal communication.

Index

■ ■ ■ ■ ■

The Authors

Dr Iván Bócsa (left) and Michael Karus (right)

Professor Iván Bócsa, born in 1926, is the academic adviser at the GATE Agricultural Research Institute in Kompolt, Hungary, where he has been active in breeding and cultivating hemp for more than forty years. He also holds a chair at the Agricultural University at Godollo in Hungary.

Michael Karus, born in 1956, is an academically trained physicist. As managing director of the nova Institute for innovation and ecology in Hürth, he heads the largest hemp research department in Germany. The nova Institute has also produced the first international hemp conferences: the Bioresource Hemp Symposiums of 1995 and 1997. Karus is a leading force behind the rediscovery of hemp in the German-speaking countries.

If your favorite retailer or catalog is sold out of our hemp books, use the order form below. For quantity discounts, please contact us.

FAX ORDERS: (419) 281-6883

TELEPHONE ORDERS: Call TOLL-FREE 24 hours: (800) 265-4367 or (419) 281-1802. Credit cards accepted.

POSTAL ORDERS: Make check payable to Bookmasters*, 1444 US Route 42, Ashland, OH 44805 USA

INQUIRIES: Call (707) 823-2800 ■ www.hemptech.com

* Fulfillment house for HEMPTECH

Qty	Publication	Price	Total
	The Cultivation of Hemp: Botany, Varieties, Cultivation and Harvesting – by Dr. Iván Bócsa and Michael Karus	$18.95	
	Hemp Horizons: The Comeback of the World's Most Promising Plant – by John Roulac, HEMPTECH founder and president	$18.95	

Shipping and handling: $5 for the first book and $3 for each additional book in US, $7 Canada & Mexico, $10 elsewhere.

Sales tax: Please add 7.5% for books shipped to California addresses, and 6% for books shipped to Ohio.

Total enclosed

Payment: ❏ Check ❏ Visa ❏ MasterCard ❏ American Express

Credit card number _____

Name on card _____ Exp. _____

Ship to:

Name _____

Company _____

Address _____

City _____ State _____ Zip _____

Telephone _____

■ ■